Green Energy and Technology

For further volumes:
http://www.springer.com/series/8059

Milan Janić

Greening Airports

Advanced Technology and Operations

Springer

Milan Janić
Department of Transport and Planning
Faculty of Civil Engineering and Geosciences
Delft University of Technology
Stevinweg 1
2628 CN Delft
The Netherlands
e-mail: M.Janic@tudelft.nl

ISSN 1865-3529 e-ISSN 1865-3537
ISBN 978-0-85729-657-3 e-ISBN 978-0-85729-658-0
DOI 10.1007/978-0-85729-658-0
Springer London Dordrecht Heidelberg New York

British Library Cataloguing in Publication Data
A catalogue record for this book is available from the British Library

Cover design: eStudio Calamar, Berlin/Figueres

Printed on acid-free paper

Springer is part of Springer Science+Business Media (www.springer.com)

For my brother Predrag—a prominent architect

Preface

This book describes the potential of advanced technologies and operations for "greening", i.e., ensuring the sustainable development, of airports as a component of the air transport system, which also includes airlines and the Air Traffic Control (ATC) system. In such context, "greening" implies medium- to long-term mitigating of the airports' environmental and social impacts under conditions of their continuous constrained or unconstrained growth. For such a purpose, a range of advanced available and/or forthcoming concepts and strategies, technologies, and operational procedures and rules, are considered. The concepts and strategies include a framework, content, and methodology for monitoring, analysing, and assessing the level of greening, i.e., sustainable development of the entire air transport system and particularly one of its components—airports. The technologies, operational procedures and rules relate exclusively to airports.

Specifically, in the airport airside area, these technologies deal with assessing the prospective effects of developing particular large congested airports into true multimodal transport nodes by connecting them to the High Speed Rail (HSR) network, introducing advanced Air Traffic Control/Air Traffic Management (ATC/ATM) technologies that support innovative operational procedures and rules, which are expected to increase the airside (runway) landing capacity, and the gradual introduction of Liquid Hydrogen (LH_2) as an alternative aviation fuel. Whereas the former two points are expected to postpone the need for airport spatial expansion by building new infrastructure (runways) which sometimes requires the taking of substantial areas of new land, the latter is expected to mitigate and even diminish the consumption of crude oil as a non-renewable energy source and the related emissions of greenhouse gases by the entire air transport system and particularly airports thanks to the chemical and burning characteristics of LH_2 as an alternative aviation fuel.

In the airport landside area, the Light Rail Rapid Transit (LRRT) system is an advanced technology which is expected to improve the capacity, efficiency, and effectiveness of the ground accessibility of airports while at the same time mitigate the overall environmental and social impacts of the airport ground access systems (modes) such as noise, energy consumption and related emissions of greenhouse gases, traffic congestion, and traffic incidents/accidents. This is expected to be achieved by competitiveness of the LRRT system taking over air passengers and airport employees from other airport ground access systems/modes (mainly single-passenger cars).

The assessment of the contribution of the above-mentioned advanced technologies to "greening", i.e., ensuring the sustainable medium- to long-term development of airports and consequently of the entire air transport system, is based on the "what-if" scenario approach. This implies that almost all technologies and operations considered have either already been developed and are waiting for real-life implementation, or further elaboration is needed before being considered for such potential implementation. In any case, before being implemented, they have to be evaluated with respect to their overall social and economic feasibility, which will require taking into account their other important economic and social attributes, and particularly their safety and security.

As the author, I hope that this book will also be used as an additional source of inspiration and motivation for intensifying research on making airports greener and thus friendlier for us all, particularly for our children and grand-children, who will surely wish to fly more than their parents or grandparents.

Milan Janić

Contents

Abbreviations

A/C	Aircraft
ACI	Airport Council International
APT	Air Passenger Transport
APU	Auxiliary Power Unit
ATC	Air Traffic Control
ATM	Air Traffic Management
ATM	Air Tansport Movement
dB(A)	Decibels
EU	European Union
FL	Flight Level
GDP	Gross Domestic Product
GPS	Global Positioning System
HSR	High Speed Rail
ICAO	International Civil Aviation Organization
IFR	Instrument Flight Rules
ILS	Instrumental Landing System
IMC	Instrument Meteorological Conditions
K	Kelvin Degree
LCC	Low Cost Carriers
LRRT	Light Rail Rapid Transit (system)
LTO	Landing and Take-off (cycle)
MLS	Microwave Landing System
MTOW	Maximum Take-off Weight
p-km	Passenger Kilometres
RPK	Revenue Passenger Kilometre
RTK	Revenue Ton Kilometre
RWY	Runway
SFC	Specific Fuel Consumption
TOW	Take-off Weight

t-km	Ton-kilometres
VFR	Visual Flight Rules
VMC	Visual Meteorological Conditions

Chapter 1
Introduction

The greener or more sustainable development of the air transport system implies its continuous growth, contributing to a range of social-economic effects (benefits) while simultaneously mitigating or even diminishing its negative impacts-costs on the environment and society in both relative and absolute terms. The former mainly includes the system's contribution to the local, regional, and national GDP (Gross Domestic Product) through employment, trade, related welfare, and globalisation of economies and societies. The latter includes consumption of energy obtained from non-renewable sources (fossil fuels–crude oil) and the related emissions of greenhouse gases, particularly CO_2 (Carbon Dioxide) and NO_x (Nitrogen Oxides), local surrounding noise and land use by airports, congestion and delays of airlines and their users—air passengers and cargo shipments, air traffic incidents/accidents, waste, security of personnel and service users, safety of operations, and related activities.

In Europe in particular, the objectives related to diminishing the above-mentioned impacts (costs) are to be achieved by pioneering and exploring advanced technologies and operations, as stated in EU policy documents such as The Vision 2020 and the EU Transport White Paper. The objectives include:

1. Developing ATC/ATM (Air Traffic Control/Air Traffic Management) advanced operations around airports and in airspace, supported by existing and/or advanced technologies, which will increase the efficiency and effectiveness of aircraft flights in terms of time, cost, and fuel consumption on the one hand, and improve the level of safety on the other; in addition, they are expected to increase the airport runway capacity and consequently mitigate the need for building new infrastructure requiring additional land;
2. Developing advanced aircraft designs in terms of propulsion, lift, and both exterior and interior shape, which all are expected to contribute to improving the overall flight efficiency;
3. Introducing advanced aviation fuels required by the development of advanced propulsion technologies, which could substantively reduce the energy consumption of existing fuels obtained from non-renewable sources (fossil fuels–crude oil),

M. Janić, *Greening Airports*, Green Energy and Technology,
DOI: 10.1007/978-0-85729-658-0_1,

improve efficiency of flights, and consequently contribute to diminishing greenhouse gas emissions; and
4. Developing some large congested airports into true multimodal transport nodes and designing and constructing advanced off-shore airports, which are both expected to mitigate impacts such as energy consumption and local emissions of greenhouse gases, noise, and congestion, as well as to contribute to diminishing the need to acquire new, usually scarce land for expanding airport airside and landside infrastructure.

Similarly, analogous endeavours to improve efficiency, effectiveness, safety/security, and social-environmental performance either directly or indirectly in all above-mentioned aspects are taking place in the USA (United States of America), where the large national research and development program the Next Generation Air Transport System (NextGen) is well underway.

This book focuses on some of the above-mentioned advanced technologies and operations to make the future air transport system and particularly its ground infrastructure components (airports) "greener", i.e. more sustainable in the medium-to long-term. In most cases, the "what-if" scenario approach is used, respecting the fact that particular advanced technologies and operations are still not in place but have been intensively studied at the conceptual and experimental level by academics, researchers, the air transport industry, and policy makers. In such context, true multimodalism at airports appears to be an exception, particularly in Europe, where some large congested airports have already been developed into true multimodal transport nodes, while others are still expecting such re-development.

Therefore, in addition to this introductory Chap. 1, the book consists of seven other chapters.

Specifically, Chap. 2 deals with the concept of "greening", i.e. ensuring the sustainable development of the air transport system. In addition to generally elaborating the framework, strategies, and content of greening, this chapter particularly focuses on the air transport system's energy consumption and related emissions of greenhouse gases.

Chapter 3 elaborates the methodology for monitoring, analysing, and assessing the medium- to long-term impacts of a given airport on the environment and society. Some of the impacts considered include airside congestion delays and related costs, noise, energy consumption and, related local emissions of greenhouse gases, i.e. air pollution, waste, land use (take), and air traffic incidents/accidents at airports. Such methodology enables, in addition to following the general trend(s), identification of the most important factors, causes, and, related measures for influencing these trend(s), namely improving the further process of greening.

Chapter 4 considers greening of a given airport by developing it into a true multimodal transport node. This implies connecting the given airport, which already inherently operates as a multimodal transport node at the local–regional scale thanks to its ground access systems/modes, to the long-distance global HSR (High Speed Rail) network. This has become common at many large congested airports in Europe. Under such circumstances, the HSR could replace some, particularly short-haul flights, relieve airport airside congestion, delays, related airline and air passenger costs, as well as mitigate noise and particularly the energy consumption and related local emissions of greenhouse gases. In addition, the need for building new runway(s) requiring additional land is significantly decreased, albeit only temporarily.

Chapters 5, 6 investigate greening the airport airside area by introducing advanced technologies and operations. Specifically, Chap. 5 analyses increasing the landing capacity of airport runway(s), which could be achieved through innovative operational procedures and rules supported by advanced technologies. In this case, in addition to directly contributing to reducing airside (runway) congestion and delays of airlines and air passengers, such increase in runway capacity will indirectly decrease the need for building new runway(s) and the consequent use of additional land, albeit at least temporarily. Chapter 6 investigates diminishing fuel consumption and related emissions of greenhouse gases at both the global level (air transport system) and the local level (at a given airport), which could be achieved in the long-term by gradually replacing conventional Jet A fuel (kerosene as a derivative of non-renewable crude oil) with an alternative fuel—renewable LH_2 (Liquid Hydrogen). In this respect, developing cryogenic propulsion technologies (aircraft engines) is highlighted as an essential challenge.

Chapter 7 considers the potential contribution to greening the airport landside area by introducing an advanced LRRT (Light Rail Rapid Transit) system for ground accessibility. In the medium- to long-term, the LRRT system is expected, in addition to providing additional transport capacity and thus improving the effectiveness and efficiency of the overall airport ground accessibility, to contribute to mitigating the overall externalities of the airport ground access systems such as road congestion and related time losses of air passengers and airport employees, energy consumption and associated emissions of greenhouse gases, noise, and traffic incidents/accidents. Specifically, the LRRT system's potential in mitigating energy consumption and related emissions of greenhouse gases, coupled with the introduction of electric cars as a prospective forthcoming airport ground access alternative, is investigated.

The final Chap. 8 summarises some of the conclusions.

Each chapter is based on the author's previous research and organised and structured as a scientific paper, consisting of the following sections: introduction, description of the system and problem to be dealt with, a methodology for analysing and assessing the potential contribution of particular advanced technology and/or operation to greening a given airport, application of the methodology to the selected airport, and some concluding remarks. Each chapter ends with a list of sources for further reading.

Chapter 2
Greening the Air Transport System: Structure, Concept, and Principles

2.1 Introduction

Over the past two decades, greening (ensuring the sustainable development of the air transport system) has been considered as an important part of the agenda by almost all the system's involved parties. These include: (1) aviation organisations for international cooperation; (2) international aviation organisations; (3) air transport system operators such as airports, ATC (Air Traffic Control/management), and airlines; (4) aerospace manufacturers of aircraft, engines, avionics, and other supportive facilities equipment; (5) non-governmental organisations and lobby groups; (6) users such as air passengers and air cargo shippers; and (7) research, scientific and consultancy organisations [11, 19]. Despite the diversity of involved parties and their interests, in most cases, the concept of greening or sustainable development has been viewed, until recently, rather narrowly by emphasising eco-efficiency, usually considering only a few types of impacts—energy consumption and related local and global emissions of greenhouse gases, and local noise around airports. More recently, other impacts such as land use (take) by the air transport infrastructure (mainly airports and their connection to the catchment area) and particularly congestion and delays have been taken into account. At the same time, the costs of these impacts in terms of the environmental and social damages have been elaborated, but not yet systematically internalised.

This chapter describes the structure of the air transport system as a part of the entire transport system and the concept of its prospective medium to long-term greening, i.e., ensuring its more sustainable development on a global scale. The structure is described by the relevant technological and operational characteristics of the particular system components-airlines, ATC/Management, and particularly airports. The concept of "greening" is elaborated through the system's contribution to mitigating emissions of greenhouse gases on the global scale [19]. Thus, this chapter serves as an introduction to elaborating greening of airports with advanced technologies and operations described in the forthcoming chapters.

M. Janić, *Greening Airports*, Green Energy and Technology,
DOI: 10.1007/978-0-85729-658-0_2, © Springer-Verlag London Limited 2011

2.2 Structure of the Air Transport System

2.2.1 *Background*

In general, similar to other transport modes, the air transport mode (system) consists of two major components: demand and supply (capacity). Demand includes sub-components such as users—air passengers and air freight (cargo) shipments. Supply (capacity) components include resources needed to be deployed to carry out the air transport services, such as labour, capital, and energy allocated to three main sub-components-airlines, airports, and ATC/management—all operating according to dedicated national and international regulations. In this context, airports and ATC are considered as air transport system infrastructure, providing the space, facilities, and equipment for servicing air users-passengers, freight (cargo), and airlines (aircraft), which represent the demand for this infrastructure. Regarding the relationship with users (passengers and freight shippers), airlines provide aircraft as 'mobile infrastructure' to serve them. The airlines, airports, and ATC use facilities and equipment supplied by the aerospace manufacturers according to the given institutional (national and international) regulations aiming to guarantee safe, efficient, and effective operations.

2.2.2 *Airlines*

Individual airlines and their alliances constitute the airline industry. In general, they operate different air route networks in terms of the spatial configuration, number of routes, flight frequency, and aircraft fleets. An airline's technical/technological performances are mainly influenced by the characteristics of the aircraft fleet it uses. Materialised through aircraft design, these characteristics include aircraft speed, carrying capacity, and productivity. The airline operational performance includes demand, capacity, and quality of services. In this section some technical/technological and operational characteristics of airlines/aircraft relevant for greening the air transport system and consequently airports are described.

2.2.2.1 Technical/Technological Performance

General

The main feature of the air transport system that has enabled its global development is the high speed of the means of transport (aircraft) as compared to that of other transport modes. In addition, the aircraft carrying capacity (payload) and maximum take-off weight have increased together with speed. Specifically, aircraft

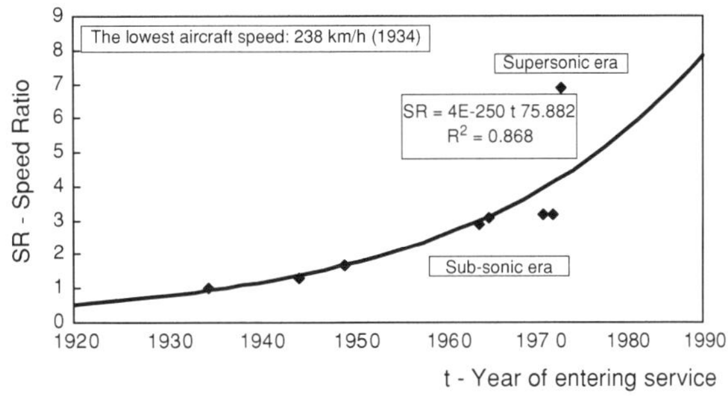

Fig. 2.1 Development of aircraft speed over time (Compiled from [19])

speed has increased over time, peaking at the development of the supersonic aircraft—Concorde—in 1974 [4]. This trend is shown in Fig. 2.1 by comparing the speed of the various commercial aircraft with the speed of the slowest aircraft-DC3 (238 km/h).

As can be seen in Fig. 2.1 above, the trend indicates a disproportional increase of aircraft speed over time, currently peaking with development of the supersonic aircraft—Concorde—in 1974 [4].

The aircraft payload has also increased in line with increased diversity of aircraft types. For example, today's aircraft fleet ranges from small (e.g. 30–50 seats) aircraft to the most recent giant Airbus A380 with a seating capacity of 500–600 seats. The speed-payload development has certainly contributed to an increase in aircraft technical static and dynamic productivity—the former defined as the available payload depending on the range and the latter defined as the product of a given payload and corresponding block (i.e., gate-to-gate speed). Figure 2.2 shows these performances for different aircraft types [19].

As can be seen in Fig. 2.2a, the maximum payload for a given aircraft type can be carried up to a certain distance. After that, the range can be increased only at the expense of the payload. Figure 2.2b shows that at the same time, technical productivity increases thanks to the increasing aircraft flight speed with range (distance). However after reduction of the payload in order to increase the range, it decreases despite continuous increase in flight speed. In addition, Fig. 2.2b shows that the substantive increase in technical productivity of new generation aircraft has been achieved over the past two decades, as indicated by A300, B757, B767 300ER, the anticipated B787, and particularly A 380 aircraft. The latest aircraft is quite superior in terms of both static and dynamic technical productivity as compared to other aircraft in the given example [19].

Aircraft Engines

The above-mentioned increase in aircraft technical productivity and the consequent coverage of relatively long distances in a relatively short time, have been mainly possible thanks to the development of high bypass turbofan engines

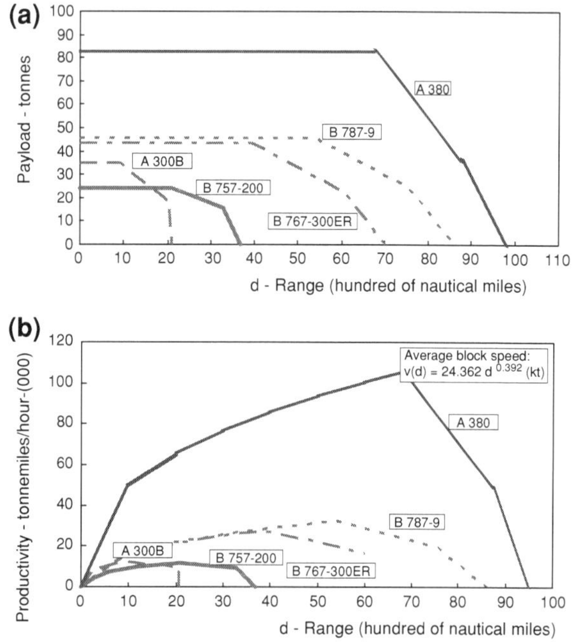

Fig. 2.2 Aircraft production performances depending on stage length (Compiled from [18]). **a** Payload vs. range. **b** Productivity vs. range

[19, 20]. These engines consume a derivative of crude oil-Jet A (or JP1) fuel (kerosene), with engine thrust, efficiency, SFC (Specific Fuel Consumption), emissions of greenhouse gases, and noise as their most significant operational, economic, environmental, and social parameters, respectively.

Engine thrust is derived from the change in momentum of the air passing through the engine and the thrust occurring due to the static pressure ratio across the final (exhaust) nozzle. Analytically, this is expressed as [19, 20]:

$$T = m(v_1 - v_0)/g + (p - p_0)/A \tag{2.1}$$

where

m is air flow through the engine (kg/s);
v_1 is the velocity of exhaust jet (m/s);
v_0 is the velocity of air entering the engine (m/s);
g is the gravitational acceleration (m/s^2);
p, p_0 is the pressure at the intake and exhaust station, respectively, (N/m);
A is the nozzle cross sectional area (m^2)

In expression 2.1, the thrust T is usually expressed in kilo-Newton (kN) (SI units) or Libras (lb) (British units).

Engine efficiency (η_e) directly expresses the rationale of engine fuel consumption, i.e., higher efficiency implies lower fuel consumption per unit of engine thrust.

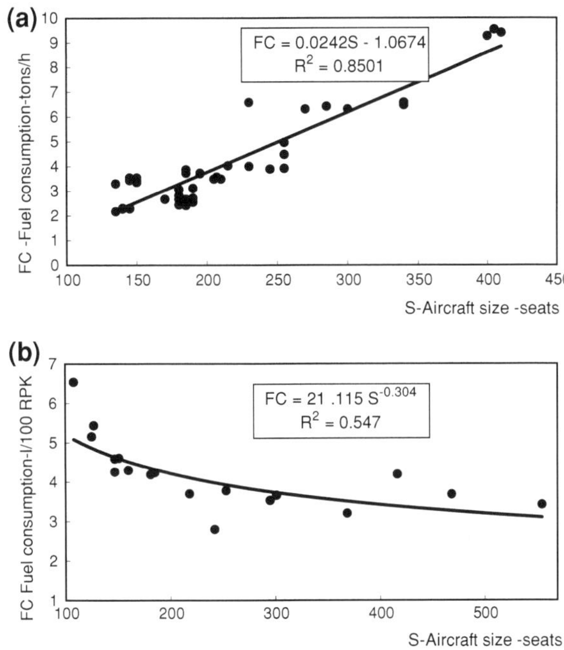

Fig. 2.3 Aircraft fuel consumption depending on seat capacity (Compiled from [18]). **a** Quantity per unit of time—tons/h. **b** Quantity per unit of output-l/RPK

SFC (*Specific Fuel Consumption*) expresses the amount of fuel generating one unit of thrust in a given unit of time. It is expressed in kg of fuel per kN of thrust per h. The SPC and engine efficiency η_e are interrelated as follows:

$$\mathrm{SFC} = \frac{M}{4\eta_e} \tag{2.2}$$

where

M is Mach number

For example, the SFC for most contemporary turbofan engines with an efficiency of around $\eta_e = 35\text{–}40\%$ operating above the tropopause at $M = 0.84$ is about 0.55–0.60 [19, 20]. Multiplying SFC with the thrust per engine and the number of engines per aircraft may give an estimation of the total fuel consumption per unit of time of a given flight. For example, this consumption increases with the aircraft size (i.e., seat capacity) and the number of engines, implying that larger aircraft have higher total engine thrust. Figure 2.3a shows an example of this relationship.

As can be seen, fuel consumption increases almost linearly with the aircraft size (seating capacity). On the other hand, Fig. 2.3b shows that the average fuel consumption per unit of output (in this case expressed in litres of fuel per 100 p-km) decreases disproportionally with increase of the aircraft size (seating capacity). For the commercial aircraft used in Fig. 2.3, fuel consumption represents a significant

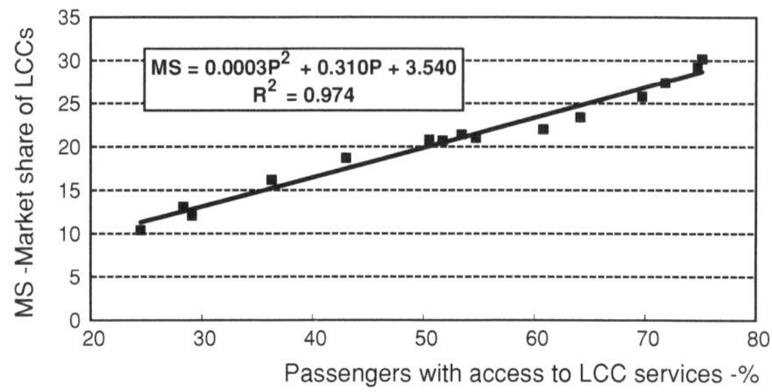

Fig. 2.4 Development of LCCs (Low Cost Carriers) in the US domestic market—period 1998–2007 (Compiled from [1])

portion (around 30%) of direct operating costs. Therefore, fuel-efficient engines play an important role in the overall economic efficiency of the aircraft and airline. Indirectly, lower fuel consumption per single unit (engine) implies lower consumption of non-renewable energy sources and consequently lower emissions of air pollutants.

Emissions of Greenhouse Gases

Emissions of greenhouse gases from aircraft engines in terms of type and quantity of air pollutants depend on the engine bypass ratio and fuel type. The bypass ratio of modern turbofan engines is around 10. As mentioned above, these engines consume Jet A (or JP1) jet fuel, of which the most known products, directly proportional to the quantity of fuel burnt, are CO_2 (Carbon Dioxide) (3.18 $kgCO_2$/kg of fuel) and H_2O (water vapour) (1.2 kgH_2O/kg of fuel). Thus, longer flights consume larger quantities of fuel and consequently emit larger quantities of CO_2 and H_2O over relatively large geographical areas (intercontinental flights are a typical example).

2.2.2.2 Operational Performance

Demand, capacity, and quality of service can be considered the main components of airline operational performance.

Demand

Airline demand is usually expressed by the number of transported passengers and/or the volume of freight, as well as by the amount of revenue passenger (p-km) and/or the revenue from ton-kilometres (miles) carried during a given period of time (day, month, year). In both cases, one revenue p-km or t-km, respectively, implies one passenger or one tone of cargo carried over the distance of 1 km (instead of km, statute miles are also frequently used: 1 mile = 1.609 km).

In some cases, the p-km and t-km revenue figures are aggregated after converting approximately ten revenue p-km into one revenue t-km. In general, these volumes increase simultaneously with any or both influencing factors. In order to realise a large p-km volume on short haul distances, the airline should carry out a larger number of shorter flights, providing passenger demand justifies this. The number of longer distance flights, again justified by passenger demand, can be lower in order to realise the equivalent volume of revenue p-km. In addition to volume, airline demand depends on time (seasonality and monthly, weekly, daily peak, and off-peak periods), and directionality. The latter is a common characteristic of air freight (cargo) demand. Airline demand is usually forecasted for the short-and long-term. Short-term forecasting can be done daily, weekly, or monthly, while long-term forecasting usually encompasses one or more years. Forecasting can be carried out for a single airline, airline alliances, or the airline industry as a whole. In terms of the geographical scale of the market, forecasting can relate to a single airline route, part or the whole airline or its alliance's network, or to the entire airline industry of a given country or region (continent).

Capacity

The airline fleet usually consists of different aircraft types characterised by the seat-and cargo-volume capacity. Large full-cost or network (legacy) airlines usually operate large (global) networks consisting of routes of different lengths and passenger and freight volumes, which require deployment of different aircraft types in order to efficiently and effectively serve such diverse demands. In general, the smaller-regional, medium narrow-bodied, and wide-bodied, as well as large narrow-bodied and wide-bodied aircraft serve regional short, continental-medium, and intercontinental long-haul routes, respectively [2].

In general, low-cost airlines or LCCs (Low Cost Carriers) usually operate in the short- and medium-haul markets using uniform aircraft fleets consisting mostly of a single aircraft type-in most cases from the B737 and/or A319/320 aircraft families. Despite operating such unified fleet on short-to medium-distance routes, these airlines have grown tremendously mainly in domestic O–D (Origin–Destination) markets. Figure 2.4 shows an example of such development in the US domestic air transport market for the period 1998–2008.

As can be seen, during the observed period (1998–2004), the market share of LCCs continuously increased from about 10% in 1998 to about 30% in 2007 thanks to increasing the accessibility of their services from about 24.5% to about 75%. The term "accessibility of service" implies the percentage of passengers with access to LCCs' low fares [1].

Airlines assign their fleets to serve the expected demand in their networks. The product of the number of seats and covered distances (i.e., length of route) represents the airline output (i.e., capacity) expressed by the volume of the aircraft seat-or ton-kilometres (s-km and t-km, respectively) carried out during a given period of time (day, week, month, year). In terms of served demand, airline output is expressed by revenue p-km. The available s-km and/or t-km are related to the p-km and t-km by the load factor, actually reflecting the percentage of the airline output (capacity) that is sold, i.e., utilised [13, 18]. The load factor is always less than 100% which enables absorption of short-term fluctuations (increases) in demand. The load factor can also be used to specify the flight frequency on a given route [3, 18].

In general, airlines operating larger networks with longer route block times and more frequent flights need to deploy a greater number of aircraft. Since flight frequency depends on demand, the size of an airline fleet will also directly depend on the volume of the expected network demand, as well as on the capacity and load factor of each aircraft. In addition, if the average utilisation of an aircraft during the season and/or year is higher, the number of aircraft required will be lower, and vice versa. The size of the airline fleet can be readjusted after determining the departure and arrival times of particular flights, and consequently the detailed itinerary of each aircraft in the network [3].

2.2.3 Airports

Airports are part of the air transport system infrastructure. They can be of different sizes, depending on the volume of traffic they accommodate in terms of air passengers, aircraft movements (atm), and air cargo during a given period of time (hour, day, year). Generally, each airport consists of an airside and landside area. The most important physical attribute is airport size, reflecting the area of land taken for infrastructure such as runways, taxiways, the apron/gate complex, passenger terminal and cargo building(s), and the ground access systems, respectively. In addition, fixed, semi-mobile and mobile facilities, equipment, and devices provide services to users—aircraft, passengers, and cargo shipments. Infrastructure, service facilities, and equipment are characterised by the service-processing rate, i.e., capacity, which depends on their constructive characteristics and users' service rules and procedures. The total installed capacity depends on the volume-time pattern of demand for service over a given period of time.

2.2.3.1 Infrastructural and Technical/Technological Performance

Spatial Layout

The main infrastructural characteristic of an airport is its size measured by the area of land it occupies. This area of occupied land depends on the airport layout and the size of particular airside and landside components. The size of particular components is governed by standards mainly related to the configuration and the number of runways, taxiways, apron/gate complex, and related facilities and equipment. These standards, specified as recommendations and design rules, depend on the size of the relevant ("critical" or "largest") aircraft and the expected volume of atm, passenger, and cargo demand [15]. In addition, the number and orientation of runways-the most land/space demanding components given the "critical" aircraft-depend on the required usability factor of the given airport with respect to the prevailing weather (wind and ceiling) conditions, which should not be less than 95% over the year. Furthermore, runways should be positioned in a way which ensures that approach and departure areas are free of obstacles on the one hand and sufficiently far from populated areas to minimise aircraft noise on the other [15, 19].

In addition, each airport is physically connected with its catchment area by various surface transport access systems such as road-based buses, taxis, and cars, and rail-based trains. All these systems use fixed dedicated infrastructure (roads, railway lines). In particular, rail-based systems can be fairly diverse in terms of the scope of connectivity, from those connecting a given airport to its catchment area (i.e., local connectivity) to those connecting the airport to the national and international rail network(s) (i.e., global connectivity). The size of an airport's catchment area differs at various airports and is usually measured by either the accessibility distance and/or the accessibility time for a given percentage of users—air passengers and/or air freight shipments. For example, in most European airports, these range between 15–100 km and 0.5–2 h, respectively [9].

Facilities and Equipment

In addition to the fixed infrastructure, each airport is equipped with various fixed, semi-and fully mobile facilities and equipment needed to handle aircraft (airside area), air passengers, and air freight (landside area). In the airport airside area, two categories of facilities and equipment are generally used. The first category encompasses fixed components, such as lighting systems near and on the runways, taxiways, and aprons enabling the aircrafts' smooth operation (landing, taking-off, and taxiing) under low visibility conditions (darkness, dense fog, very low cloud, rain, and snow storms). The navigational aids enabling approaching and departing the airport are not taken into account even though the lighting system might be a part of them. The second category consists of vehicles serving aircraft entering and leaving the apron parking gates (stands) and those providing aircraft servicing during the turnaround time (e.g. refuelling, catering, cargo, waste, vehicles for delivering and collecting passengers and their baggage, and power supply vehicles).

In the airport landside area, mobile and semi-mobile "interfaces" enable physical connection of the aircraft to the passenger terminal(s). For example, passengers may use airport buses (mobile units) in combination with mobile stairs and/or air-bridges (semi-mobile facilities) to pass from the terminal building to the aircraft, and vice versa. Passenger (and freight) terminals enable air passengers and freight, respectively, to transfer from the airport ground access systems to the aircraft, and vice versa, i.e., to change transport mode. In such contexts, the terminals operated by the particular airport ground access modes could be integrated with the airport passenger terminal where various facilities and equipment are used for such transfer, i.e., processing passengers and/or freight. The airport ground access systems include individual cars and taxicabs, and mass transport systems such as buses, and regional and national conventional and HS (High Speed)–rail. At many large airports, all the above-mentioned ground access systems provide transport services to air passengers, airport employees, and other visitors.

2.2.3.2 Operational Performance

Demand

Demand is one of the most important planning and operational parameters of a given airport. It is expressed by the volume of requests for services during a given period of time (hour, day, month, and year). These requests come from atms (air transport movements), passengers, and freight shipments (one atm is equivalent to one landing or one taking-off).

Demand in the airport airside area in terms of the number of atm from past periods could be used as an indicator for planning and managing airport airside capacity. For example, the number of atm per hour (or 15 min intervals) is usually used for specifying aircraft/flight delays given the airport's ultimate capacity, and consequently for declaring the airport's practical capacity. The latter usually amounts to about 80–85% of the former.

Demand, expressed by the number of air passengers to be handled during a given period of time, is usually considered as one of the basic inputs for planning, designing, and operating the airport landside area, i.e., components such as the passenger terminal complex, interfaces, and the ground access systems. For example, the current and forecasted annual number of passengers, after being converted into hourly peak passenger volumes, can be used for sizing the passenger terminal complex and its particular components, as well as for determining the capacity of passenger processing facilities. These include ticketing/check-in, immigration and body check counters, baggage claim devices and surrounding areas, etc. The conversion of annual into hourly volumes of passengers can be carried out using different methods discussed in this book further below. In addition, details of the prospective airline schedules, aircraft fleet mix, and the load factor can also be used as design parameter(s) more precisely. In such contexts, categorisation of passengers can be important since different categories place different demands on the various components of the passenger terminal complex at different times. For example, all passengers can be broadly classified as originating, terminating or transit/transfer, then as domestic or international, and finally as business and leisure passengers. At hub airports, there is usually a high proportion of transfer/transit passengers. In addition, some airports are specialised for serving charter or LCC traffic. Such increased complexity in estimating the relevant volumes and structure of demand as sizing parameters for both airport airside and landside areas has been partially resolved by using the scenario approach and determining the relevant peak-hour demand [19].

Capacity

Capacity is usually defined as the maximum number of units of demand that can be accommodated during a given period of time under given conditions. In the airport airside area, air transport movements (atms) represent demand (one atm is either one landing or one take-off). In the airport landside area, the volumes of air passengers and/or freight shipments represent demand. Airport capacity (providing there are no other constraining factors) mostly depends on operational factors such as 'safety constraints', 'constant demand' for service, and 'average delay' per unit of accumulated demand. This capacity, if used for operational purposes, is usually determined for each hour, including an average delay per operation, while for planning purposes, it is determined for the period of one year. However, in many cases, different economic and environmental constraints may affect the airport operational capacity. In such cases, the concept of airport economic and/or environmental capacity is introduced. The airport economic capacity is mainly dictated by short-and long-term economic constraints. In the short-term, these might represent fees for airport services during the peak and off-peak periods aiming to regulate airport access, covering the increased costs of service and reflecting the type of users and their willingness to pay for service. Such fees should also be compatible with the ICAO (International Civil Aviation Organization) recommendations and bilateral airspace agreements. In the long-term, investment availability for airport expansion usually determines the economic conditions and thus the prospective long-term capacity. The airport environmental capacity takes into account the environmental constraints in terms of noise and allowed emissions of greenhouse gases set up to protect the local population and the environment from adverse effects. In the short-term, this capacity is expressed similar to operational capacity regarding the above-mentioned environmental constraints. In the long-term, land take constraints may compromise the airport's spatial expansion, its capacity, and consequently its growth [6, 8].

The capacity of the airside area includes the runway system, taxiways, and apron/gate complex. The capacities of these components should be balanced in order to avoid "bottlenecks" and consequent adverse effects such as airline and air passenger congestion, delays, and related costs.

The capacity of the airport landside area includes the capacity of the passenger and freight terminal complex. For the passenger terminal complex, the capacity of particular components regarding their basic function can be determined as follows [18]:

- Processors, i.e., passenger and baggage servers;
- Reservoirs, i.e., waiting areas for passengers (including people accompanying them) and their baggage; and

- Links, i.e., areas equipped with facilities and devices connecting processors and reservoirs.

Processors serve passengers on their way from the airport ground access systems to the aircraft and vice versa. Reservoirs provide space for air passenger queuing and waiting for particular phases of the service process. Links include long corridors, passageways, walkways, and escalators connecting particular processors and reservoirs. Passenger baggage is dealt with in parallel. Specifically, departing passengers deliver their baggage at check-in counters. Such baggage proceeds to the baggage sorting area where it is semi-or fully automatically sorted on a flight by flight basis and then delivered to the aircraft. Arriving baggage is delivered directly from the aircraft to the baggage claim area where it is picked up by arriving passengers from moving, usually rotating baggage claim devices. In general, two concepts of the ultimate capacity of the passenger terminal complex and its particular components can be considered. The "static" ultimate capacity implies the maximum number of passengers (occupants) in an area of a given size, in such a manner that each occupant is provided with a minimum area. On the other hand, the "dynamic" capacity implies the maximum processing/service rate of a given service facility. Each passenger can be given the maximum waiting time for service [18].

The capacity of the airport ground access systems is determined by the number of seats offered during a given period of time to air passengers and airport employees. Specifically, for public road-and rail-based systems, this capacity depends on the service frequency and vehicle size per frequency [18].

2.2.4 Air Traffic Control/Air Traffic Management

The ATC/ATM (Air Traffic Control/Air Traffic Management) is the third crucial component of the air transport system. It consists of the controlled airspace over countries, continents and oceans, radio-navigational facilities and equipment located on the ground and in space (satellites) and their complements on board aircraft, operating staff (air traffic controllers), and operating rules and procedures for safe, efficient and effective guidance of each individual aircraft, as well as air traffic flows. The ATC/ATM around airports controls/manages incoming and outgoing aircraft traffic and thus contributes to the efficient and effective use of the available airport airside infrastructure. In other words, the airport airside capacity crucially depends on the ATC/ATM safety separation rules and their ability to deliver the aircraft safely in the shortest possible time slots, which consequently enables the maximum planned number of atms (air transport movements) during a given period of time. In addition, in order to maintain the required level of safety and prevent overloading of the particular airspace and airport components, the ATC/ATM imposes delays on particular atms and thus more or less directly (airside) and indirectly (landside) influences the quality of service provided to users—airlines, air passengers, and air cargo shippers at any given airport.

In the context of airspace around airports, the technical/technological performance includes the characteristics of existing and new radio-navigational facilities and equipment, airspace organisation, and aircraft separation rules. The relevant operational performances include parameters such as demand, capacity, and their matching on the operational, tactical, and strategic levels. The latter implies the safe, efficient, and effective movement of individual aircraft and air traffic flows through a given controlled airspace.

2.2.4.1 Physical and Technical/Technological Performance

Organisation of Airspace and Aircraft Separation

The ATC/ATM is established over a given airspace to provide safe, efficient, and effective guidance of air traffic (aircraft/flights). Safety requires the respect of air traffic separation rules and the serving of users—aircraft/flights—without creating potential conflicts. Efficiency and effectiveness are concerned with providing aircraft/flights smooth movement along the fuel-cost optimal trajectories connecting the origin and destination airports without deviations due to ATC/ATM reasons.

Meeting the above-mentioned requirements (objectives) requires division (organisation) of the controlled airspace into smaller parts depending on traffic intensity (density) and complexity. Such division is carried out by: (1) dividing the airspace into airport zones, terminal areas, low and high altitude en-route areas; and (2) dividing each of these areas into the smaller parts-'ATC/ATM sectors', each under the jurisdiction and responsibility of one or a team of air traffic controllers. In particular, airport zones enable management of the arriving and departing traffic flows. Terminal airspace established above airport zones enables managing more intensive and complex arriving and departing traffic flows. This airspace covers the area with a radius of about 40–50 nautical miles (1 nm = 1.852 km) around any given airport. Usually, this area begins vertically at the ground level up to FL (Flight Level) 100 (each flight level represents a constant altitude of 10^3 ft (1 ft $\sim$ 0.305 m)). The aircraft fly through this area along the prescribed arrival and departure trajectories defined by the radio-navigational facilities/equipment and/or ATC radar vectors [14].

In the airport zone and terminal airspace, each aircraft can fly according to IFR (Instrument Flight Rules) or VFR (Visual Flight Rules). VFR flights can take place exclusively under so-called VMC (Visual Meteorological Conditions), while IFR flights are carried out under both VMC and IMC (Instrument Meteorological Conditions). For example, in Europe, only IFR is exclusively applied under both IMC and VMC, while in the US, both VFR and IFR are applied depending on weather conditions. IFR aircraft/flights are primarily responsible for maintaining the assigned flight paths while the ATC/ATM maintains the separation rules, whereas VFR aircraft/flights perform primary navigation and maintain the prescribed separation rules between other VFR and IFR aircraft/flights. In any case, they remain under the jurisdiction and monitoring of the ATC/ATM. In an airspace shared by both IFR and VFR flights, the division of responsibility between the pilots and the air traffic controllers for maintaining safe separation is always carried out according to the IFR. Current research programs concerning the modernisation of ATC/ATM system both in the US (NextGen) and Europe (SESAR) aim to improve on-board and ground navigation capabilities and consequently to decentralise the existing ATC/ATM system by allocating more responsibility for aircraft separation and the operational air traffic management from ATC controllers to pilots.

Facilities and Equipment

The ATC/ATM can use different existing and advanced technologies (facilities and equipment) for monitoring and controlling individual aircraft and air traffic flows in the airspace between airports, in the vicinity of airports, and at the airports themselves. In order to achieve safe, efficient, and effective "gate-to-gate" air transport operations, the following particular ATC/ATM components should be considered indivisible:

- Communication facilities and equipment consist of the communication channels for transmission of information between pilots and air traffic controller(s) (VHF/UHF air/ground voice and non-voice communication links); the communication links established between particular ATC/ATM control units; the communication links providing the exchange of information between the ATC/ATM and other parties involved in air traffic control. One promising communications system is VHF data link (VDL Mode 2 in the medium- and VDL Mode 3 and 4 in the long-term). This data link is used to automatically (without voice) communicate a wide range of flight parameters from the aircraft to the ground (air traffic controller). This substantially improves air traffic controllers' awareness of the situation, reduces overall (particularly communications) workload, and consequently increases system capacity, efficiency, and effectiveness. In addition, Controller Pilot Data Link Communications (CPDLC) using data link for ATC/ATM communications, D-FIS (Data Link Flight Information Service) enabling pilots to receive the flight information through the air/ground data communications, and satellite data-link enabling air/ground communication in cases when an aircraft is out of range of the ground communication systems, have also been developed [10, 21];
- The navigational facilities and equipment include ground aids, airspace satellites, and corresponding devices on-board the aircraft. Depending on their location and primary function in the terminal airspace and airport zone, they can be classified into several groups, as follows: (1) the external overland terminal airspace aids (VOR, DME, VOR/DME, ILS-(Instrumental Landing System, MLS-Microwave Landing System, and ADS-B SLS-Satellite Landing System),which use data from the GPS (Global Positioning System) and/or

GLONAS (GLObal NAvigation Satellite System), (2) the airport external navigational equipment (approach lighting systems, slope indicators, surface detection equipment); and (3) internal overland terminal aids (RNAV systems) [10, 18, 21]. Contemporary navigational procedures will be primarily based on global navigation satellite (RNAV) systems providing aircraft with direct 4D (Four Dimensional) trajectories between their origin and destination airports rather than using the airway system, which does not always follow the shortest-great circle distance. Such 4D trajectory routing will enable more flexible utilisation of the available airspace and consequently increase the possibility of flying along preferred fuel-cost routes;
- The surveillance facilities and equipment utilise different radar systems. Two radar types are available: primary and secondary (beacon) radar SSR (Secondary Surveillance Radar and Mode S). They allow ATC controllers to monitor traffic on the radar screen and to distinguish between aircraft [10, 21]. In advanced ATC/ATM, surveillance capability will be significantly improved by ADS-B (Automatic Dependent Surveillance Broadcast) system as a complement to SSR for the ATC and primary traffic monitoring tool for aircraft. In particular, ADS-B in combination with CDTI (Cockpit Display of Traffic Information) will enable pilots to monitor and keep safe distance from surrounding traffic on the one hand and periodically broadcast information obout their position and the other flight-relevant parameters to the ATC on the other. Implementation of both ADS-B and CDTI will require precise division of responsibilities in executing particular ATC/ATM tasks (particularly those concerning aircraft separation) between pilots and ATC controllers. In addition, A-SMGCS (Surface Movement Ground Communication System) has been developed to enable surveillance, guidance, and routing of aircraft on the ground under all local weather conditions;
- Decision support tools include Arrival and Departure Management System (AMS and DMS respectively), which, as ground-based traffic management automation tools, optimize incoming and outgoing aircraft flows at a given airport. They are expected to relieve congestion by improving utilisation of the runway system capacity. The runway management system complements these tools and both should reduce the workloads of air traffic controllers and consequently increase system capacity.

The above-mentioned technologies will enable introduction of advanced operational procedures such as CDA (Continuous Descent Approach), Steeper Approach procedure, procedures with displaced landing threshold to mitigate noise, and eventually time-based separation rules instead of the current distance-based separation rules for landing aircraft. As described in Chap. 5, the former will make flights more fuel-efficient and less noisy in the vicinity of airports, while the latter will increase airport capacity and consequently—at least temporarily—reduce the need for building new runways to efficiently and effectively handle the growing airport demand.

2.2.4.2 Operational Performance

Demand

The demand at a given airport in terms of the number of atms per given period of time represents also the demand for the corresponding ATC/ATM unit. The hourly number of flights is relevant for operational purposes and short-term prediction of air traffic controllers' workload, while for long-term strategic handling of demand, the annual number of flights is relevant. This number is also important for planning the long-term development of ATC/ATM capacity [12].

Capacity

The capacity of a given ATC/ATM sector (unit) including that around airports is usually expressed by the maximum number of aircraft/flights served during a given period of time under given conditions. These conditions are specified regarding: (1) the size and configuration of the aircraft approach and departure paths around a given airport; (2) the facilities and equipment available to both ATC/ATM and the aircraft, and (3) the intensity, volume, structure, time and space distribution, and continuity of demand. The ATC/ATM capacity includes the capacity of particular components such as the airspace around a given airport, air traffic controllers, and communication links [18, 22]. In order to estimate the capacity of a given ATC/ATM sector based on the estimation of air traffic controller workload, the number, duration, and order of execution of particular control tasks (activities) performed for the air traffic (aircraft/flights) passing through a given sector according to a given pattern during a given period of time should be considered [18].

2.3 Concept of Greening the Air Transport System

2.3.1 Background

The above-mentioned characteristics of the air transport system, in addition to the obvious benefits to society in terms of contribution to the GDP (Gross Domestic Product) and local and global welfare for particular involved parties, create impacts on the environment and society. In general, the main impacts include energy consumption, emissions of greenhouse gases contributing to climate change, local noise around airports, land use by airports, air and ground traffic congestion and airline and air passenger delays, and air traffic incidents/accidents. Expressed in monetary terms, these impacts are considered externalities [19]. In the present context, greening the air transport system implies stabilizing and/or diminishing all these impacts (externalities) in both absolute (total quantities) and relative (quantity per unit of output) terms under conditions of continuous growth of air traffic volumes. In this respect, system greening through advanced technologies and operations is focused on potentially reducing the direct energy consumption and consequent emissions of greenhouse gases as matters of global concern (the other aspects of greening, i.e., more sustainable development at the local—airport scale are addressed in the forthcoming chapters).

Despite the total quantities being relatively small compared to those from other transport modes and man-made industrial activities, CO_2 (Carbon Dioxide) emissions of the air transport system are specific mainly due to the fact that they are almost immediately deposited at altitudes between 9 and 12 km (below and above the tropopause) and latitudes from 40°N to 60°N, where, together with NO_x (Nitrogen Oxides) emissions, they act as greenhouse gases contributing to global warming. In order to mitigate such impacts, quantitative targets of such emissions have been set. For example, in Europe, the ACARE (Advisory Council for Aeronautical Research in Europe) has set targets for improving the environmental performance of the whole air transport system, covering aircraft engines, airframe, and operations. These targets imply reduction of the fuel consumption and CO_2 emissions (units per passenger-kilometre) by about 50% by the year 2020 (of which about 10–20% relate to contributions from improving engines and the rest from improving airframe and operational efficiency). During the same period, emissions of NO_x (Nitrogen Oxides) are expected to reduce by 50%, of which about 60–80% will be achieved by engine improvements and the rest by improvements in airframe and operations. However, these improvements, if achieved, will only slow the increase in the cumulative emissions of greenhouse gases which will occur if air passenger and freight transport demand continue to grow. Under such circumstances, a reasonable solution is to change the fuel, as described in Chap. 6 [19, 23].

2.3.2 *Aircraft Engine Fuel Consumption*

Raising awareness of the depletion of crude oil reserves and the consequently potentially limited availability of jet A fuel and its derivatives at given prices, as well as of the harmful impacts of burning such fuel on public health and the environment, has stimulated aerospace manufacturers to continually improve the fuel efficiency of existing jet engines and consequently contribute to mitigating emissions of greenhouse gases. The design of such engines has consequently embraced solving a range of complex problems, the most complex being balancing the engines' propulsion and thermal efficiency. Better propulsive efficiency has provided a greater propulsive power from the combustion process while improved thermal efficiency has generated a higher overall engine pressure ratio and turbine temperature using the same amount of fuel (energy). Other problems relate to the proper balancing of engine weight, drag, noise, and emissions ratios.

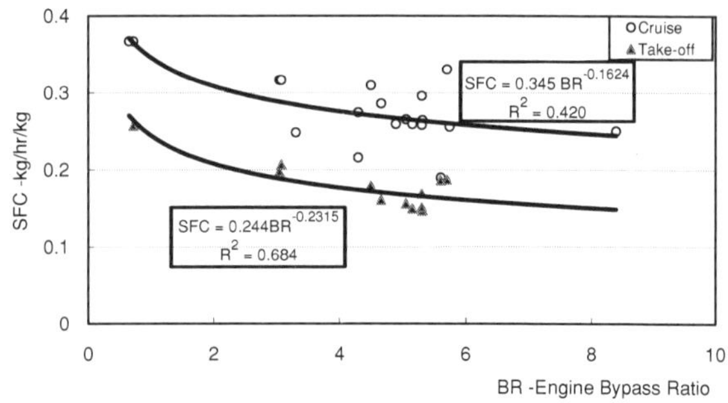

Fig. 2.5 SFC (Specific Fuel Consumption) vs. BR (Bypass Ratio) for jet engines (Compiled from [19])

In order to obtain a higher propulsive efficiency it has been necessary to reduce waste energy in the engine exhaust stream, which has decreased jet velocity. Since engine thrust is the product of the exhaustive mass flow and its velocity, if this velocity was reduced, the mass flow would be increased to retain the desired level of thrust. This implies an increase in the bypass ratio defined as the rate between the amount of air flowing around the engine core and the amount of air passing through the engine itself. Engines with the higher bypass ratios usually have lower Specific Fuel Consumption (SFC), defined as the ratio of fuel burned per hour per tonne of net thrust [17, 19]. The SFC of most contemporary jet aircraft engines amounts to about 0.25–0.30 kg of fuel/kg of thrust/hour. This is likely to be the case until around the year 2015, when further reductions of up to 0.184 kg of fuel per hour per kg of thrust will occur. Specific Fuel Consumption (SFC) relates to the jet engine bypass ratio (BR). The nature of this relationship is illustrated using data for 20 engine types produced by the airspace manufacturers such as CFM Company (joint corporation of Snecma (France) and General Electric Company (USA)), Rolls-Royce (UK), Pratt & Whitney and General Electric (USA), and IAE (International Aero Engines AG made up of the engine manufacturers Pratt & Whitney, Rolls-Royce, MTU (Europe) and Aero Engine Corporation (Japan)) The regression relationship, in which the bypass ratio (BR) is considered as the independent and SFC as the dependent variable, is illustrated in Fig. 2.5.

As can be seen, SFC (Specific Fuel Consumption) has, independently of the flight phase (take-off or cruising), decreased more than proportionally with the engine bypass ratio, which might be useful for estimating the development trends of commercial aircraft jet engines.

Improvements in aerodynamic performances have also played an important role in the improvement of aircraft fuel efficiency. An illustration of this is the development of the most recent Boeing B777-300ER (ER—Extended Range). In case of this aircraft, the more fuel-efficient "raked wing tip" design has replaced the winglets option used previously on other B777 versions as well as on the B-747-400 and B737 NG (Next Generation) aircraft. The new winglets are expected to improve the short-field climb performances and fuel efficiency by about 1–2% on longer flights. If the aircraft is typically utilised for about 5,000 h per year with average fuel consumption of about 7.5 t/h, a fleet of 20 aircraft with an average depreciation period of about 20 years will consume about 15 million tons of fuel. Thus, a fuel saving of around 1% equates to 150 thousand tons, consequently producing 477 thousand fewer tons of CO_2 and about 177 thousand fewer tons of H_2O (water vapour) [19]. In addition, combined with improvements in ATC/ATM operations expected to be achieved through current research programs such as NextGen (US) and SESAR (Europe), the relative fuel efficiency of the airline industries in these regions is also expected to improve. Figure 2.6 shows the possible trend for the US airline industry.

As can be seen, during the period 1960–2005, the fuel consumption in kg per ton-kilometre decreased more than two-fold from about 0.92 kg/t-km to about 0.43 kg/t-km as the industry's output, expressed by the annual volume of ton-kilometres, increased. Some forecasts for the next decade (2010–2017) indicate that the average fuel consumption will remain at the present level (0.42/043 kg/t-km) despite further growth of volumes of system output. However, some doubts on further improving fuel efficiency in the above-mentioned context have remained. Namely, the question has been raised whether the system has already or will soon exhaust its potential for further improvements in fuel efficiency of about 2.5–3% until and beyond the year 2017/2020 while using the same or the very similar but improved jet engines technology (see also Fig. 2.3) [12, 19].

2.3.3 Emissions of Greenhouse Gases

Emissions of greenhouse gases from the air transport system are considered as air pollutants. These substances appear in much higher concentrations than in the natural environment (i.e., the hypothetical environment without human influence). As such they may damage people's health and the environment (flora and fauna). In general, the main greenhouse gases emitted by burning Jet-A aviation fuel by commercial aircraft are CO_2 (Carbon Dioxide) and H_2O (water vapour), NO (Nitric Oxide) and NO_2 (Nitrogen Dioxide), which together form NO_x (Nitrogen Oxides), SO_x (Sulphur Oxides), and smoke. The emission rates of CO_2, H_2O, and SO_2 are relatively constant −3.18 kg/kg of fuel, 1.23 kg/kg of fuel, and up to 0.84 g/kg of fuel, respectively. The emission rate of NO_x changes i.e., increases with the pressure ratio of the jet engine, which in turn increases the jet engine's thermal efficiency. The engine pressure ratio is defined as the ratio of the total pressure at the compressor discharge and at the compressor entry. For contemporary turbofan engines, this ratio ranges from 10 to 50, which originates from the typical design of the combustion chambers of these engines [19]. Experiments to investigate the relationship between the engine emission rate of NO_x, compressor outlet temperature, and pressure ratio have resulted in a regression equation as follows [19]:

$$ER_{NO_x} = 0.17282\, e^{0.00676593 T_s} \tag{2.3}$$

where:

ER_{NOx} is the engine emission rate of NO_x (g*NOx*/kg of Jet A fuel); and

T_s is the compressor outlet temperature ranging between 280 and 1,080 K (degrees Kelvin)

Expression (2.3) indicates that the emission rate of NO_x increases with the compressor outlet temperature, albeit at diminishing rate. In addition, a description of characteristics of the above-mentioned greenhouse gases and their perceived impacts on the environment follows.

2.3.3.1 CO (Carbon Monoxide) and CO_2 (Carbon Dioxides)

CO (Carbon Monoxide) is always produced during the burning of fossil fuels of which Jet A fuel is a derivative. It reacts with oxygen (O_2) in the atmosphere and forms Carbon Dioxide CO_2. The emission rate of CO_2 from Jet A fuel is almost constant −3.18 kgCO_2/kg of fuel, which makes it easier to estimate the quantities emitted at both local and global levels based on the quantities of fuel burnt. Emissions of CO_2 have a long lifetime in the atmosphere (about 100/150 years). There is no remedy for reducing the unit quantity of CO_2 emissions by improving the fuel burning process in existing jet engines simply because of the fuel chemistry.

2.3.3.2 H_2O (water vapour): Contrails

H_2O (water vapour) emitted after burning Jet A fuel influences climate change through the formation of contrails in the troposphere (10–12 km), the cruising height of most commercial aircraft. Contrails are the icy clouds formed behind an aircraft flying at high altitudes; they are often visible from the Earth's surface during clear skies. They form as follows: behind an aircraft, warm exhaust gas containing particles of soot, ash, and other pollutants expands and mixes with the colder and dryer air. If the amount of water in the air is at saturation level, water droplets form. Due to the low ambient temperature (at altitudes of about 10–12 km, the temperature is about −40°C or lower), these droplets rapidly freeze and form ice crystals, which under conditions of sufficient water vapour build up very quickly into persistent and visible contrails (clouds). The layer of the atmosphere where this process occurs is called the "contrails producing layer". Some estimates have shown that contrails cover about 0.1% of the Earth's surface and thus contribute to the Earth's overall coverage by high clouds.

2.3.3.3 NO_x (Nitrogen Oxides)

The symbol NO_x implies nitrogen oxides NO and NO_2 ($NO_x = NO + NO_2$). They are produced during any combustion in which air in the form of N_2 and O_2 is brought to a high temperature by burning fuel. NO_x is formed in the flame at a temperature of a few thousand K (degrees Kelvin) and generally its formation increases together with the burning temperature. Jet A fuel burns at such temperatures, while the second source is the Jet A fuel itself, containing about 1% of NO_x. The remedy for NO_x is generally twofold: (1) reducing the fuel-burning temperature, which generally reduces the jet engine fuel-burning efficiency; and (2) lowering the available oxygen for combustion. Commercial jet engines generate NO_x according to Eq. 2.3. The amount is greatest during take-off and climbing when the engine burning temperature is at its highest, slightly lower during cruising, and lowest during the approach and landing phases of flight.

2.3.3.4 SO_x (Sulphur Oxides)

Crude oil and its derivative—jet aviation fuel-may contain considerable amounts of sulphur. In a chemical reaction with the water vapour in the atmosphere, acid rain is created, damaging trees and other dependent natural habitats. Catalysts are added to fuel to diminish the presence of sulphur in jet engine fuel and exhaust gases.

2.3.3.5 NMHCs (Non-Methane Hydrocarbons)

HCs (Hydrocarbons) contribute to the formation of smog and global warming. However, the amounts emitted by burning jet fuel have not been recognised as particularly worrying as compared to other types of air pollutants. Nevertheless, the contribution of HCs to global warming appears to be relatively important through (1) the production of ozone O_3, (2) extending the lifetime of methane CH_4, and (3) their conversion into CO_2 and H_2O, the most important greenhouse gases.

2.3.4 Impact on Global Warming and Climate Change

2.3.4.1 Physics

Generally, life on Earth is dependent on the physical properties of the Sun–Earth system. The surface temperature of the Sun is about 5,800 K, which results in an emission spectrum with a maximum wavelength of 500 nm* (nm*—nanometre; 1 nm* = 10^{-9} m). This gives a solar temperature of the exact magnitude to induce photochemical reactions. Depending on the radius of the Sun and the Earth, their distance, and the above-mentioned surface solar temperature, one can estimate that the earth receives energy of 1,379 W/m^2 (Watts per square meter), although the solar constant is always taken to be slightly lower, i.e., $S \approx 1{,}370$ W/m^2. Supported by some gases in the Earth's atmosphere, this energy appears sufficient to maintain an average temperature on the Earth's surface of $T = 288$ K (+15°C). A part of the received energy is reflected from the Earth's surface back into space. This is known as albedo from the Latin term "albus" meaning "white".

Astronomers usually use albedo to express the brightness of the Earth as seen from space. Consequently, the energy equation can be set as follows: $(1 - a)\pi R^2 S = 4\pi R^2 \sigma T^4$, where R is the Earth's radius (6,400 km) and σ is the Stefan–Boltzmann constant ($\sigma = 5.672 \times 10^{-8}$ Wm^{-2} K^{-4}). With an estimated value of albedo of $a = 0.34$, one can obtain the temperature of the Earth's atmosphere of $T = 250$ K (usually it is taken to be $T = 255$ K). This is lower than the Earth's surface temperature (288 K), which is mainly due to the presence of gases such as CO_2, O_3, NO_2, CH_4 (Methane), and H_2O. Otherwise, this temperature would be lower by about 30 K. In general, these gases absorb most of the heat radiation from the Earth and reemit it back towards the Earth's surface, a process called the "greenhouse effect". As a result, these gases are referred to as "greenhouse gases". Currently, the concentration of greenhouse gases is continuously increasing due to both natural and human causes. For example, the concentration of CO_2 has increased by about 25% over the past 200 years; the level of CH_4 has doubled during the last 100 years, while the concentration of NO_x has been increasing by about 0.25% per year. The air transport system (flights) has contributed to this.

Increasing the concentration of greenhouse gases may strongly influence the climate by increasing the average temperature on the Earth's surface. During the past 130 years, the average global temperature has increased by about 0.6 K. The speed and scope of the process is still not precisely known. For example, according to past estimates, the air transport system contributed to about 2% of the total man-made emissions of CO_2 in 1992. This is a small but significant contribution to the total impacts of "greenhouse gases". CO_2 has a very long residence time in the atmosphere, where it mixes well with other gases. Some simple estimates can show that, for example, an instant doubling of the concentration of CO_2 relative to the present concentration would increase the average temperature on the Earth's surface by about 1.4 K. This phenomenon can be explained as follows: increasing the concentration of CO_2 will reduce the Earth's long wavelength radiation in the top layers of the atmosphere by a certain amount and consequently reduce the inward flux there by the same amount. The energy balance in the top layers of the atmosphere requires a constant flux. Therefore, the Earth's surface temperature must rise in order to compensate for this imbalance. This effect is called radiative forcing. Some estimates have shown that the air transport system might contribute to increasing radiative forcing by about 0.02 W/m^2. In general, any increase in the global temperature may cause additional effects—increasing or mitigation of the concentration of CO_2 as a reversible process. Some estimates suggest that the current concentration of CO_2 in the Earth's atmosphere is around 382 ppm (ppm-parts per million) and this is likely to increase at an annual rate of about 1.2 ppm over the next 40 years, that is by the year 2050. Other estimates indicate that when the total known crude oil reserves of about 1,650 billion (10^{12}) U.S. barrels are exhausted (currently expected by the end of the twenty first century), the concentration of CO_2 will contribute to the increasing of the average global temperature by about 2.5 K.

An equally important gas in the Earth's atmosphere is ozone (O_3). Its presence protects the Earth from harmful solar UV radiation by absorbing all light with a wavelength of less than 295 nm* (nano-meter). The layer of O_3 in the earth's atmospheres is relatively thin-about 0.3–0.4 cm—and is under a constant temperature and atmospheric pressure. Although the gas is present throughout the atmosphere, its highest concentration is in the stratosphere at altitudes of about 20–26 km from the Earth's surface. Ozone is constantly formed through reaction of molecular oxygen O_2 and atomic oxygen O influenced by solar UV radiation. Most ozone is formed above the equator where the amount of UV solar radiation is at its highest. From there, it moves towards the poles where it is "accumulated" up to a thickness of about 0.4 cm during the winter period [19].

However, ozone is sensitive to free radicals such as atomic chlorine CI, nitric oxide NO, and hydroxyl radicals OH, which are formed from water vapour (H_2O) and chlorofluorocarbons (CFCs), products of burning aviation fuel, which escape from the troposphere (10–12 km from the Earth's surface) where most commercial flights take place to the stratosphere where the ozone layer is formed. At these altitudes, free radicals, including NO_x, lead to depletion of the ozone layer. Those that do not escape remain extremely stable in the troposphere where they, together with NO_x, contribute to thickening of the ozone layer. The residence time of NO_x in these regions increases with altitude. Therefore, NO_x affects the ozone layer regionally if injected into the troposphere and globally if injected into the stratosphere. In any case, the increased concentration of the above-mentioned pollutants might generally cause depletion of the ozone layer with inevitable impacts. For example, depletion of this layer by about 10% may cause an increase in the UV radiation by about 45%, which certainly inflicts damage to almost all biological cells and in particular causes skin cancer in those persons who expose their skin to sunlight.

The gaseous H_2O emitted from burning jet A fuel contributes to forming clouds both at high-altitudes and nearer the ground. In particular, clouds near the Earth's surface affect the atmosphere by reducing the amount of solar radiation returning to space. High altitude clouds (contrails) contribute to an increase in the amount of solar radiation reflected from the atmosphere. Consequently, the surface becomes warmer in order to keep the radiative forces in balance. Some estimates indicate that contrails contribute to radiative forcing by 0.007–0.06 W/m^2 with the expectation that they will increase with the projected air traffic growth to about 0.04–0.4 W/m^2 by the year 2050 [16].

2.3.4.2 The Scale of Impact

The impact of the air transport system to climate change, i.e., global warming, could be roughly estimated by using a zero-dimensional greenhouse model [19]. The model is based on considering the total energy flux at the top of the atmosphere. Under conditions of equilibrium, the radiation flux vanishes, i.e., inward and outward radiation are in balance. This balance can be disrupted by a reduction in the Earth's long-wave radiation at the top of the atmosphere by an amount equal to that caused by the increase in greenhouse gases (for example, CO_2). Consequently, the outward radiation will decrease by the same amount. Since the energy balance at the top of the atmosphere requires a constant flux, the temperature at the Earth's surface will increase by the amount of ΔT_s in order to compensate the reduction of the Earth's long-wave radiation ΔI. This phenomenon, known as radiative forcing, puts variables ΔI and ΔT into the following relationship [19]:

$$\Delta I = \frac{\partial I}{\partial T_s}\Delta T_s \text{ where } \partial I/\partial T_s = 4/T_s(1-a)S/4 \tag{2.4}$$

Where:

- a is the albedo of the Earth as viewed from space ($a = 0.34$; otherwise, at the Earth's surface, $a = 0.11$);
- T_s is the Earth's surface temperature ($T_s = 288$ K);
- S is the solar constant ($S = 1.370 \times 10^3$ J/sm^2 (Joules per second per square meter))
- I is the radiative force (W/m^2)

The common values of particular parameters gives an estimate of expression (2.4) of: $\partial I/\partial T_s = 3.1\,\text{W/m}^2\,\text{K}$ and its reciprocal value of: $G = 0.32\,\text{m}^2\text{K/W}$. In addition, from Eq. 2.4 it follows that: $\Delta T_s = G\Delta I$. The numerous models of climate change specify the values of radiative forcing of about ΔI = 4–4.6 W/m^2 as the contribution of the man-made emissions. Some reports suggest that the air transport system might contribute to this total up to a maximum of 3.5%, i.e., its radiative forcing would be: $\Delta I_a = 0.035\,\Delta I$ = 0.14–0.16 W/m^2. Applying this to the equation $\Delta T_s = G\Delta I$ gives $\Delta T_{s/a} = G\Delta I_a$ = 0.3(0.14–0.16) = (0.042–0.048) K. Another input on parameter ΔI suggests that air transport could contribute to an increase in the Earth's surface temperature by $\Delta T_{s/a}$= (0.052–0.096) K between the year 2010 and 2050 [16, 19]. Both results suggest that the Earth's surface temperature will not significantly increase due to the air transport system operating under the given circumstances. In other words, the air transport system seems not to significantly contribute to global warming. However, more investigation is needed to confirm or reject this hypothesis particularly due to the high sensitivity of the available models to input data, which is itself the output of other very complex climate change models. Therefore, the problem of greening, i.e., more sustainable development of air transport system in terms of energy consumption, consequent emissions of greenhouse gases and contribution to global warming and climate change remains high on the agenda of all parties involved in its future medium-to long-term development.

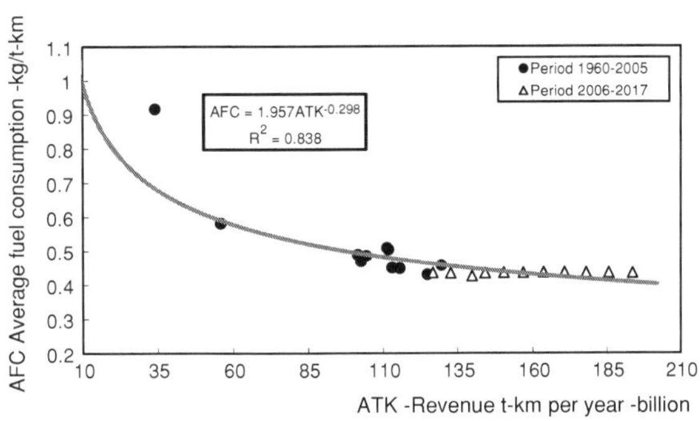

Fig. 2.6 Fuel efficiency vs. output volume for the US airline industry (1960–2017) (Compiled from [5, 12, 19])

2.4 Principles of Greening the Air Transport System

The main principles of greening, i.e., more sustainable development of the air transport system in terms of its energy consumption, emissions of greenhouse gases, and other impacts on the environment and society can be summarized as follows:

"Continuing the medium-to long-term growth driven by internal and external socio-economic driving forces while simultaneously maintaining and/or diminishing the cumulative global impacts on society and the environment". These two inherently opposite developments make achieving this greening principle extremely complex despite substantive achievements in diminishing the impacts in relative terms, i.e., per unit of system output (see for example Fig. 2.6). Namely, the world's air transport system has grown from 0.5 trillion RPKs (Revenue Passenger Kilometres) in 1971 to about 4.85 trillion RPKs in 2008. Some long-term forecasts of international air transport organizations (IATA, ICAO, ACI), and in particular by the two main manufacturers of commercial aircraft (Boeing and Airbus), predict continuous and rather stable growth at an average annual rate of 5% over the period of the next 20 years, which will increase the volume of RPKs to about 10.545 trillion RPKs or 11.4 trillion RPKs by the year 2025/2026. At the same time, the number of passengers is predicted to grow at an average annual rate of 4.5%, which will result in their total number of about 6.8 billion by 2025/2026. In addition, air cargo traffic is forecasted to grow at an average annual rate of 6.1% over the same period, from about 200 billion RTKs in the year 2006 to about 650 billion RTKs in the year 2025/2026. The above-mentioned air traffic growth will require an increasing number of aircraft, from the current 18,230 (of which 16,250 are passenger aircraft) in 2006 to about 36,420 (of which 32,440 will be passenger aircraft) by 2025/2026. Since all these aircraft continue to be powered by conventional Jet A fuel as a derivative of crude oil, the cumulative fuel consumption and related emissions of greenhouse gases CO, CO_2, SO_4, and NO_x, will continue to increase despite the above-mentioned improvements in aircraft technology (IPCC, 1999). Some estimates indicate that the 513 $MtCO_2$ emitted by the air transport system in the year 1992 is expected to increase to about 1,468 $MtCO_2$ by

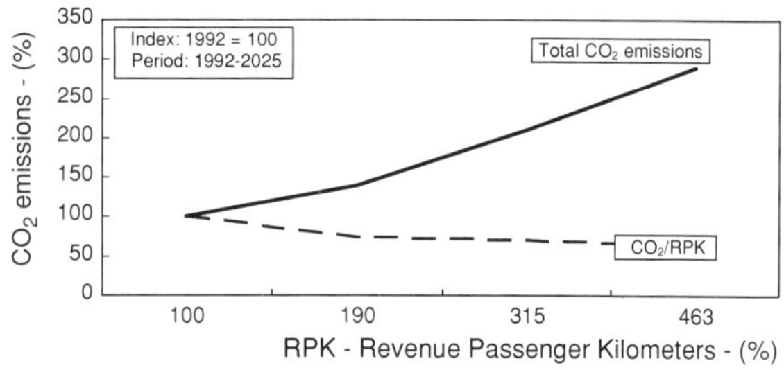

Fig. 2.7 The prospective long-term development of global CO_2 emissions from the air transport system in the EU (Compiled from [7, 19])

the year 2050. The latter quantity will likely amount to between 3 and 5.5% of the total CO_2 emissions caused by man [19]. These estimates indicate that greenhouse gas emissions by the air transport system are likely to increase about five-fold in absolute terms if no substantive improvements take place, compared to current figures. Figure 2.7 shows such possible development in the 27 Member States of the EU (European Union) [2, 7, 19].

As can be seen, in the EU, the relative quantities of greenhouse gas CO_2 per RPK will slightly decrease with the annual volume of RPKs over the forthcoming decade and a half. At the same time, the total cumulative annual CO_2 emissions will continue to increase slightly disproportionally, at an increasing rate due to the increasing volumes of traffic. This is because the manoeuvring space for influence is relatively limited under the given circumstances. In order to investigate the scope for possible influence, let the total annual emissions of CO_2 be expressed as follows [19]:

$$Q_E = pdF_cS_E \tag{2.5}$$

where:

Q_E is the total emission of greenhouse gases (tons);
p is the number of air travellers;
D is the average travel distance (km or miles);
F_c is the fuel consumption (tons per RPK); and
S_E is the specific emission or the emission rate (tons of pollutant per ton of fuel consumed)

In expression (2.5), the variable S_E for particular polluting gases such as CO_2 and H_2O will remain constant for existing Jet A fuel. The variable F_c could eventually but certainly not substantially be further decreased by the above-mentioned development of new aircraft airframes and engines, more efficient and effective maintenance, and significantly improved flight and air traffic flows management, guidance, and control. Consequently, the two variables subject to economic and regulatory measures are the number of passengers p and the average travel distance d. Reducing these two variables may prove a very complex task. Due to the increase in GDP and overall prosperity, more people will travel by air, taking advantage of constantly diminishing airfares. In addition, the globalisation of business and tourism will probably contribute to increasing average travel distances, which in turn, together with an increase in the number of people travelling, will make the volume of passenger kilometres (RPKs) grow. Under such challenging circumstances, the question remains whether it is possible at all to further make air transport greener in absolute terms. The answer is affirmative, but only by replacing current fuel with an alternative, possibly LH_2 (Liquid Hydrogen). As described in Chap. 6, this fuel will certainly be able to eliminate further growth in emissions of CO_2, but at the same time it will increase growth in H_2O emissions both at the global and local-airport scale. This latter will contribute to greening airports in parts of the air transport system and consequently the entire system as well as in terms of diminishing accumulation of greenhouse gases [19].

2.5 Concluding Remarks

Despite enormous efforts, developments so far indicate that it will be very difficult for the air transport system to move towards the above-mentioned concepts and principles of greening, i.e., more sustainable development, which implies increasing the effects (benefits) and mitigating the impacts (costs). The main reason for this is satisfying growing air transport demand efficiently, effectively, and safely. In particular, airlines have been deploying an increased number of more productive (greater and faster), fuel-efficient and less air polluting, quieter, and safer aircraft. Airports have been generally taking more sizeable areas of land for building both their airside and landside infrastructure in order to provide sufficient capacity to accommodate growing air transport demands. In some cases, they have moved yet closer to the populated area, exposing local population to increased noise burdens.

The ATC/ATM has permanently improved efficiency and effectiveness of operational procedures in airspace and especially around airports in order to improve the flight fuel and greenhouse gas emission efficiency and reduce congestion and delays. Further improvements in guiding individual flights and air traffic flows are expected after introducing more advanced technologies, operational procedures, and regulations.

Evidence so far indicates that the air transport system has been making substantive progress in greening, i.e., ensuring more sustainable development in the medium-to long-term (past) period mainly in the relative terms, i.e., by mitigating particular impacts per unit of system output (RPK and/or RTK). However, accommodating continuously and rapidly growing demand at annual rates above the rates of improvements in efficiency and effectiveness of operations and particular services, has resulted in the increase of all types of impacts on the environment and society in absolute terms, i.e., in their totals, from the system. This is particularly noticeable in the total emissions of greenhouse gases and the land utilisation by airports, both of which have been steadily increasing. Consequently, the prospective alternative solutions to mitigating these impacts will certainly be moving from existing crude oil-derived Jet A aviation fuel (kerosene) to less polluting alternatives in combination with improvements of aircraft engine fuel efficiency on the one hand and the aerodynamic characteristics of aircraft on the other. At the same time, specifically at airports, it will be necessary to introduce advanced technologies, ATC/ATM procedures, and regulations to increase their airside capacity and consequently improve their efficiency, effectiveness, and greening performance over the medium-to long-term period. The following chapters intend to address some of these issues.

References

1. ACI North America (2005) Top ten myths about us airports: the future of air transportation: Challenges and opportunities for airports. FAA forecast conference, Airport council international north america, Washington DC, USA, p 30
2. AEA (2005) European aviation industry joint position paper on emissions containment policy. Association of European Airlines, Brussels, Belgium
3. Bazargan M (2004) Airline operations and scheduling. Ashgate Publishing Limited, Aldershot, England
4. Boeing (1998) Evolution of the world fleet: time line. Boeing Aircraft Company, Seattle, USA. http://www.boeing.com
5. BTS (2001) Airline service quality performance data. US Department of Transportation, Bureau of Transportation Statistics, Office of Airline Information, Washington DC, USA
6. Caves RE, Gosling GD (1999) Strategic airport planning. Pergamon, Elsevier science, Amsterdam, The Netherlands
7. CEC (2000) Defining an environmentally sustainable transport system, commission expert group on transport and environment. Report, commission of the European communities, Brussels, Belgium
8. DETR (2002) The future development of air transport in the United Kingdom: South East. A national consultation document, Department for Transport, London, UK
9. EC (2002) Study on competition between airport and the application of state aid. European commission, Directorate-general energy and transport F, Brussels, Belgium
10. EC (2007) Reduced separation minima: gaps identification—Airport, TMA, En route. RESET project, Deliverable D 4.2, 6th Framework program, European commission, Brussels, Belgium, p 86
11. EEC (2004) Defining sustainability in the aviation sector. EUROCONTROL Experimental centre, Bretigny-sur-orge, CEDEX, France
12. FAA (2005) FAA aerospace forecasts: fiscal years 2006–2017. US Department of Transportation, Federal Aviation Administration, Office of Policy & Plans, Washington DC, USA
13. Holloway S (2003) Straight and level: practical airline economics. Ashgate Publishing Limited, Aldershot, England
14. ICAO (1996) Rules of the air and air traffic services: procedures for air navigation services. DOC 4444-RAC/501, 13 edn. International Civil Aviation Organization, Montreal, Canada
15. ICAO (2004) Aerodromes, Annex 14, international standards and recommendation practices, 4th edn. International Civil Aviation Organization, Montreal, Canada
16. IPCC (1999) Aviation and the Global Atmosphere, Intergovernmental Panel on Climate Change, Cambridge University Press, Cambridge, UK
17. Janic M (1999) Aviation and externalities: the accomplishments and problems. Transp Res-D 4:159–180

18. Janic M (2001) Analysis and modeling of air transport system: capacity, quality of services and economics. Taylor and francis, London
19. Janic M (2007) The sustainability of air transportation: a quantitative analysis and assessment. Ashgate Publishing Limited, Farnham, Surrey
20. Jenkinson LR, Simpkin P, Rhodes D (1999) Civil jet aircraft design. Arnold Publisher, London
21. Little DA (2000) Study into the potential impact of changes in technology on the development of air transport in the UK. Final report for department of the environment transport and regions (DETR), Cambridge
22. Majumdar A, Polack J (2001) Estimating capacity of Europe's airspace using a simulation model of air traffic controller workload. Transp Res Rec 1744:30–43
23. RCEP (2001) The environmental effects of civil aircraft in flight. Special report, Royal Commission of Environmental Pollution, London, England

Chapter 3
Greening Airports I: Monitoring, Analysing, and Assessing

3.1 Introduction

Greening, i.e. ensuring the sustainable medium to long term development of airports, implies its consideration as a system of interrelated components, operations, and processes, the main strategies and tactics for making airports greener (more sustainable), and elaboration of particular effects (benefits) and impacts (costs) from their medium to long term operations. The main effects (benefits) embrace employment around airports and their overall contribution to the local, regional, and national economies (GDP—Gross Domestic Product). The impacts (costs) include damages and their costs from the airport's energy consumption and related emissions of greenhouse gases, i.e. air pollution, noise, airside, and landside congestion and delays of airlines, airport ground access systems/modes and air passengers and airport employees, land utilization, and wastage. In some cases, airport-originated air traffic incidents/accidents are taken into consideration.

Both—effects (benefits) and impacts (costs)—interrelate with each other in permanent dynamic interaction. This relationship raises the question of the strategies and tactics for managing the airport's future development, particularly due to raising public awareness of their impacts (costs). In general, two sets of strategies are available. In the first set, two strategies imply exclusively mitigating the impacts (costs) in both relative and absolute terms. The former strategy aims to decrease particular impacts (costs) per unit of airport output, i.e. the quantity of aircraft, passenger, and/or cargo shipments handled during a given period of time (usually one year). The latter strategy aims to diminish the total quantity of particular impacts generated by the airport over a given period of time (again, usually one year). According to these two strategies, on the one hand, airport growth is greener (more sustainable), while on the other it may be compromised, as its related effects (benefits). In the other set, an alternative strategy is aimed at trading-off the particular effects (benefits) and impacts (costs) and ensuring that their sum is always positive and increasing over the medium to long term. This enables rather unconstrained, but generally greener (more sustainable) growth of

M. Janić, *Greening Airports*, Green Energy and Technology,
DOI: 10.1007/978-0-85729-658-0_3,

the given airport. At present, depending on the case at hand, both strategies can be applied.

This chapter describes a methodology for monitoring, analysing, and assessing the achieved level of greening (sustainable development) of a given airport. This methodology consists of the concept and strategies for greening, the most important effects (benefits) and impacts (costs), and an indicator system for their quantification. As such, the methodology can represent a component of the "tool" enabling particular parties—local and national authorities and communities (policy makers), airport operators, airlines, air passengers, and other involved parties—to monitor, analyse, assess, and consequently manage their specific contribution to the medium to long term greening (sustainable development) of their airports.

3.2 Airports and Concept of Greening

3.2.1 The Main Components and Characteristics

Every airport consists of landside and airside area. The former includes the ground transport access systems connecting the given airport to its catchment area, as well as the passenger (and freight) terminal complex, while the latter comprises the airspace around given airports called the airport zone (narrow airspace) and the airport terminal area (wider airspace), the system of runways and taxiways, and the apron/gate complex [4].

Airports as an essential component of the air transport system are usually regarded as its infrastructure. In general, each airport inherently operates as a multimodal transport node facilitating air transport and other surface transport modes. In addition to the airport operator, airlines, and the surface transport modal operators (usually road and rail), there are two other large groups of transport parties involved. With the exception of airport users—air passengers, the fourth and fifth group of parties can be the local population living in the vicinity of a given airport and the local (and sometimes national) authorities. In the given context, the relationships between the particular parties in terms of satisfying their individual interests must be balanced. Users' main interests include convenient and inexpensive, effective, and safe door-to-door (air) transport services. For the airport and other transport operators, the key interests are profitable, effective, and safe operation. The local population's main interests relate to the lowest possible noise levels and emissions of other pollutants, e.g. greenhouse gases. The main interests of the local and national authorities are increasing the direct and indirect contributions of the given airport to the local and national economies while simultaneously diminishing the overall impacts on the environment and society in terms of noise, emissions of greenhouse gases, land take (use), waste, and (air) traffic incidents/accidents.

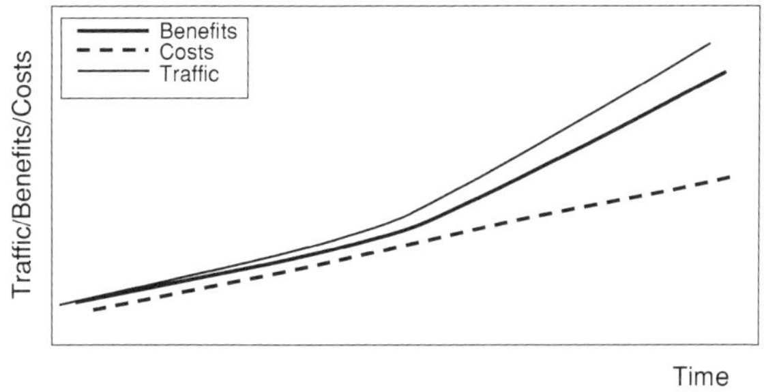

Fig. 3.1 A diagram of managing possible unconstrained greener (more sustainable) airport growth (Strategy II)

3.2.2 The Main Strategies for Greening, i.e. Sustainable Development of Airports

The main strategies of greening, i.e. sustainable development of airports, imply either continuously diminishing their impacts (costs) or continuously widening the positive gap between the overall social effects (benefits) and impacts (costs, externalities) in the medium to long term.

- *Strategy I*: Constraining airport growth. This strategy consists of constraining further expansion of the infrastructure capacity of a given airport in order to constrain the growth of air traffic and consequently maintain existing impacts (costs) mainly in terms of local noise, land take (use), and emissions of the greenhouse gases, i.e. air pollution within the prescribed limits (caps). Particular caps can be imposed either by the local and/or central authorities or possibly be inherently present factors. For example, one of the inherently present factors could be a lack of available land for expansion of the airport's infrastructure with consequent impact on its capacity. Also, land could be available but its use blocked due to high resistance from the local population. In any case, constrained growth can prevent escalation of particular impacts over the prescribed caps on the one hand, but also compromise the airport's direct and indirect positive effects (benefits) to society on the other. Consequently, the airport in question continues to operate at its existing capacity while accommodating current and eventually only slightly greater volumes of demand mainly thanks to utilising the available capacity more efficiently and effectively [35].
- *Strategy II*: Managing green growth. This seemingly the most reasonable strategy for most airports under the present circumstances consists of balancing their medium to long term growth and the related overall impacts (costs). This implies that, under conditions of growing airport traffic, effects (benefits) are maintained at the level or even above the generally increasing, stagnating and/or decreasing overall impacts (costs), both expressed in monetary terms as shown in Fig. 3.1.

3.2.3 *Effects—Benefits*

The evidence to date indicates that most airports have acted as important contributors to the local and in many cases national economies mainly in terms of direct and indirect employment, GDP, and overall welfare. The former includes the airport's staff carrying out airport-related activities in the widest sense, while the latter includes staff carrying out other activities at the airport. In general, larger airports employ more staff and thus provide a greater contribution to local and national employment, GDP, and consequently to overall welfare. This includes all direct and indirect benefits of the airport itself and its effects on the local economy. In addition, all these effects are generally directly correlated to the airport's size, i.e. the volume of traffic flows during a given period (usually one year) [28].

3.2.4 *Impacts—Costs (Externalities)*

The impacts (costs) of operation of a given airport usually include its energy consumption, local and global emissions of greenhouse gases, i.e. air pollution, local noise, congestion and delays of air passengers and airlines, land take (use), and waste. The potential impacts (costs) of air traffic incidents/accidents at airports, being rather rare events, are not considered [28]. Each of these impacts (costs) is described through its primary source and nature of impact, as an externality, and possible mitigating factor.

3.2.4.1 Energy Consumption and Emissions of Greenhouse Gases, i.e. Air Pollution

Source and Nature

Energy consumed by airports can be broadly divided into energy consumed by its airside area activities and the part consumed in its landside area activities. In the airport's airside area, this includes fuel consumed by aircraft during the LTO (Landing and Take-Off) cycles and the energy consumed by ground vehicles serving aircraft at the apron/gate complex. In the airport landside area, the main consumers of energy are the airport ground access systems/modes and passenger and cargo terminals and other administrative buildings serving the airport. In all cases, the main energy sources are non-renewable fossil fuels and in a moderate proportion also renewable wind, water, and solar sources.

Crude oil is usually used for producing Jet A fuel (kerosene), as well as other aviation fuels for powering aircraft and gasoline for other airport ground vehicles in both airside and landside areas. Electrical energy usually obtained from different sources and supplied directly to the airport through dedicated sub-stations is mainly consumed for heating, cooling, lighting, and operating the facilities, equipment, and other devices in the processes of servicing passengers and their baggage and air cargo shipments in passenger and cargo terminals, respectively. Electrical energy is also consumed for heating, cooling, and lighting other administrative buildings at the airport. Developments so far indicate that most airports have endeavoured to reduce the energy consumed per unit of their output—WLU (Workload Unit). (1 WLU = 1 passenger + his/her baggage or 100 kg of freight) [10]. For example, at Frankfurt Main airport (Germany), the average energy consumption has reduced from 16.8 kWh/WLU in the year 2003 to 14.4 kWh/WLU in the year 2008 [19].

The above-mentioned energy consumption generates emissions of greenhouse gases such as CO_2 (Carbon Dioxide), NO_x (Nitrogen Oxides), SO_2 (Sulphur Dioxide), H_2O (Water Vapour), and others. The total quantities of greenhouse gases correspond to, and can be determined separately for, the traffic-related activities in the airside area, in the landside area, and for the traffic-supporting activities. In the first case, the air pollution from arriving and departing aircraft during the LTO cycle is considered, while air pollution from vehicles servicing aircraft, passengers, and freight at the apron/gate complex can be added [28]. In the landside area, greenhouse gasses usually take the form of emissions from servicing air passengers and freight in the corresponding terminals, and from heating, cooling, and lighting these and other airport buildings. In the case of electrical energy consumption, the emissions are deemed as being indirect from sources of production. Otherwise, in the case of the following fuels: coal, crude oil derivatives, and natural gas, emissions are considered direct. In addition, both direct and indirect emissions from particular airport ground access systems/modes can be taken into account. Some measurements have shown that in relative terms, the greatest relative contributor to total emissions of greenhouse gases by an airport are the aircraft operations (LTO cycles) (about 60%) followed by ground aircraft servicing at the apron/gate complex (about 20%), the airport ground access systems/modes (15%), and electricity consumption in airport buildings (about 5%) [28].

The absolute quantities of these emissions are usually proportional to the volumes emitted, i.e. the air polluting activities and the intensity of emissions per activity. At airports, the volume of air polluting activities closely relates to the volume of air traffic usually expressed in terms of the number of passengers and aircraft movements (atms), and the weight of freight shipments (1 atm corresponds to either one arrival or one departure flight). In many cases, the number of passengers, after being expressed by their weight, and the weight of freight shipments, are converted into a common unit—WL. The intensity of emissions per air polluting activity depends on the quantity and type of energy consumed and the technology. For example, at the local airport level, if it is possible to determine the average number of activities per WLU, n_a, the energy consumption per activity e_a, the quantity of air pollution per unit of consumed energy q_{ap}, and the volume of WLUs, C_T processed during a given period of time T, the total quantity of emitted greenhouse gases will be approximately: $Q_{ap/T} \approx n_a \, e_a \, q_{ap} \, C_T^{\alpha}$ (α is an exponent).

Emissions of greenhouse gases, i.e. air pollution, at a given airport can also be considered from the global perspective. This specifically relates to airside activities. There, emissions from all incoming and outgoing aircraft/flights (atms) rather than from only the LTO cycles are considered. This implies taking into account these emissions as the consequence of maintaining a given airport in the air transport network. They can be estimated as follows:

$$Q_{ap} = \sum_{k=1}^{K} \sum_{j=1}^{M} n_k(T) t_k f_k e_{kj} \tag{3.1}$$

where

- $n_k(T)$ is the number of atms (incoming and outgoing flights) of type (k) accommodated at a given airport during the period T;
- t_k is duration of flight (k) (h) (in terms of distance, flights can generally be short, medium, or long haul);
- f_k is the average unit fuel consumption of flight (k) (ton/kg/h);
- e_{kj} is the emission rate of air pollutant (j) by flight (k) (ton of pollutant/ton of fuel/h); and
- J is the number of relevant air pollutants (these are usually CO_2, NO_x, and H_2O) [24].

The above-mentioned global emissions by a given airport would appear to be more objective especially if the air transport system and its particular components—airlines, ATC/ATM, and airports—were included in schemes of charging for externalities [2].

Externalities

The energy consumption and related emission of greenhouse gases by airports expressed in monetary terms (internalised) are considered as externalities. In particular, energy from non-renewable sources such as crude oil, natural gas, and coal actually depletes these resources and should thus be considered an externality.

As far as greenhouse gas emissions are concerned, some estimates indicate that airports contribute approximately 5% to the total quantity of greenhouse gases emitted by the air transport system, i.e. about 30 million tons per year [2]. Since this amount certainly contributes to global warming and related climate change, internalising their impact has been considered by including the international air transport system in emission trading or taxation schemes [28].

Mitigating Measures

Energy consumption and related emissions of greenhouse gases at airports can be mitigated by various operational and economic measures, which can be summarized as follows [2]:

- Reducing aircraft fuel consumption during LTO cycles and restricting the use of the APU (Auxiliary Power Unit) at the apron/gate parking stands;
- Reducing the overall number of vehicles accessing a given airport;
- Encouraging the use of low or zero emission vehicles within the airport area;
- Stimulating use of alternative fuels;
- Reducing energy consumption of all buildings;
- Including in schemes of charging for externalities, such as emission trading or taxation systems.

Operational Measures

Operational measures need to be applied to both airport airside and landside activities. In the airside area, heavily air polluting aircraft have already been banned from most airports. In addition, the share of aircraft following the ICAO (International Civil Aviation Organization) or even the improved LTO cycle needs to be increased. In the latter case, this implies shortening the taxi-in and taxi-out phase of the LTO cycles and turning the aircraft engines off for as long as possible while at the apron/gate complex. In addition, this includes optimising the movement of aircraft-servicing vehicles and other ground systems equipment at the apron/gate complex and using less-polluting existing and alternative fuels AFV (Alternative Fuel Vehicles).

In the landside area, energy used in terminal and other buildings can be reduced in both absolute and relative terms by different systems and devices for monitoring and controlling consumption. In addition, the proportion of less-polluting energy sources such as electric energy, natural gas, and particularly solar and wind energy needs to be increased. As regards airport ground access systems, increased use of more energy efficient and cleaner public transport systems/modes should be encouraged (see Chap. 7).

Economic Measures

Economic measures generally imply introducing charging schemes for emissions of greenhouse gases at airports. One such measure is air pollution tax, which airports charge per LTO cycle. The amount is set up in proportion to the fuel consumed and its emission content (for example for CO_2 it is 3.18 kg CO_2/kg of Jet A fuel) [23].

Emissions trading, implying the central authority setting a cap on the quantity of greenhouse gas emissions by a particular airport, could be an alternative charging scheme. Under such a scheme, airlines and airports are issued emission permits providing them credits on the allowable emissions within a given cap. Airports and airlines that need to pollute more can buy emission credits from those airports and airlines which have achieved savings, resulting in them using less than their credited pollution limits. Such credit transfer is referred to as a trade. The question remains how airports, as well as airlines, will be included in the scheme, in proportion to the volumes of traffic handled during the specified/agreed past period of time. This scheme is expected to be in place for all EU (European Union) internal flights by the year 2011 and for all flights in the year 2012 (http://aero-defense.ihs.com/news/en).

Both the above-mentioned charging schemes have proved unpopular at most airports, which consider them a threat to their further growth. Nevertheless, an emission trading scheme applied mainly to ground-based sources has shown to be more acceptable, despite the awareness in most airports that such schemes can also indirectly affect their growth, particularly when the achieved rates of emission reduction appear to be lower than air traffic growth [2].

3.2.4.2 Noise

The Source and Nature

The primary sources of noise at airports are arriving and departing aircraft. Different measures have been developed to quantify, monitor, and control the overall noise level. The most recent measure proposed by the EC (European Commission) measures the exposure of local population around airports to daytime and night-time aircraft noise as follows [14]:

$$L_{\mathrm{den}} = 10\,\log_{10}\left(\frac{1}{86400}\sum_{i,j}\left(N_{d/ij} + 3.16N_{e/ij} + 10N_{n/ij}\right)10^{\mathrm{SEL}_{ij}/10}\right) \tag{3.2}$$

where

$N_{d/ij}$ is the number of movements of the (j)-th aircraft group on the i-th flight path during the period on an average day;

$N_{e/ij}$ is the number of movements of the (j)-th aircraft group on the i-th flight path during the evening of an average day;

$N_{n/ij}$ is the number of movements of the (j)-th aircraft group on the i-th flight path during the night of an average day;

T_n is the duration of night (in seconds); and;

SEL_{ij} is the sound exposure level from the (j)-th aircraft group on the i-th flight path.

As with other measures, the most important characteristic is the dependence of the noise level on the number of aircraft operations during a given period of time. In addition, a measure for estimating the proportion of annoyed people close to a given airport has been developed. It embraces the proportion of those annoyed (%A) and the proportion of those highly annoyed (%HA) people from population living close to a given airport. It is dependent on the indicator of daily exposure L_{den} as follows:

$$\%A = 8.5888 \times 10^{-6}(L_{\mathrm{den}} - 37)^3 + 1.777 \times 10^{-2}(L_{\mathrm{den}} - 37)^2 + 1.221(L_{\mathrm{den}} - 37) \tag{3.3a}$$

and

$$\%HA = -9.199 \times 10^{-5}(L_{\text{den}} - 42)^3 + 3.932 \times 10^{-2}(L_{\text{den}} - 42)^2 + 0.2939(L_{\text{den}} - 42) \tag{3.3b}$$

where particular symbols are analogous to those in expression (3.2).

Externalities

In order to consider noise as an airport externality, its potential and actual damage and related costs need to be estimated. This has been achieved by using different techniques such as, for example, hedonic and contingent valuation method(s). The former is based on revealed, and the latter on stated behaviour. The hedonic price method has been most widely used for evaluation of the social cost of airport noise, at least within the academic community [6, 33]. In addition to academic efforts, charging for aircraft noise at airports has also been the subject of national and international policies. One such policy contained in documents of the European Commission proposes the following equation for charging airport noise [12]:

$$c_n = c_a 10^{\frac{L_a - T_a}{10}} + c_d 10^{\frac{L_d - T_d}{10}} \tag{3.4}$$

where

c_a, c_d is the noise charge for an arrival and a departure, respectively, which theoretically can be equal to zero (monetary units per operation);

L_a, L_d is the noise level for an aircraft at the arrival and at the departure/flyover noise certificated locations, respectively (in $dB(A)$); and

T_a, T_d is the threshold, corresponding to 95% of the total noise energy emitted at a given airport.

The main disadvantage of the method in expression (3.4) is the problem of choosing an appropriate technique to determine the values c_a and c_d, which has not yet been completely resolved. In summary, any method of charging for aircraft noise—from the cost of mitigating the noise burden to charging based on marginal social costs—can be used depending on the local circumstances and airport specifics.

Mitigating Measures

Different, rather restrictive measures have been implemented to mitigate the noise burden on local populations in the vicinity of many busy airports. Within the scope of international efforts, the 33rd ICAO Assembly 2001, introduced the concept of a "Balanced Approach" to noise management and control at airports. This implies identifying the noise problems at a given airport, analysing, and implementing mitigation measures through exploring the following elements:

- Reducing noise at source, i.e. only allowing aircraft operations according to Chaps. 3 and 4 [22];
- Restricting operation of particular aircraft types, i.e. forbidding operation of particular aircraft types during specific periods of the day;
- Using noise abatement (operational) procedures, i.e. redistributing noise, which implies use of preferential runways and approach/departure routes and noise abatement approach/landing procedures; any such procedure must satisfy the necessary safety standards;
- Planning and managing land take (use), i.e. introducing land-use zoning around a given airport aimed at minimizing the number of people affected by aircraft noise; and
- Charging excessive noise, i.e. introducing noise charges in cases where severe negative noise-related effects exist [22].

In addition, improvements in the noise performance of contemporary aircraft through investments in innovative technologies (aircraft engines) have been substantive and permanent [22].

Also, the ATC/ATM operations have developed several advanced operational procedures for mitigating noise around airports, some based on existing and others based on advanced technologies. These include LD/LP (Low Drag/Low Power), CDA (Continuous Descent Approach), IGS (Increased Glide Slope), DR (Displaced Threshold) and CA (Curved Approach) procedures [14].

3.2.4.3 Congestion and Delays

The Source and Nature

Congestion and delays of aircraft arriving at and departing from the given airport, respectively, occur whenever the intensity of demand exceeds the available airport service rate (capacity). The relationships between the various types of such occurrences may be varied. In some cases the airport's "ultimate" capacity, defined as the maximum number of aircraft/flights served during a given period under conditions of constant service demand, is overall greater than the demand rate during a given period of time. However, due to the uneven spread of demand over time, the immediate demand may exceed this capacity, causing inevitable congestion and delays of affected aircraft/flights. As the overall demand rate approaches the "ultimate" capacity, the number of cases in which the immediate demand rate exceeds the immediate service rate shall significantly increase, with the number of affected aircraft/flights and consequently the duration of their delays also rapidly increasing. Summing up the individual delays and dividing the sum by the number of all aircraft/flights demanding service during a given (busy) period of time produces the average delay per aircraft/flight. Generally, this delay directly depends on the intensity of demand on the one hand, and indirectly on the airport "ultimate" capacity (as a reciprocal of the minimum average service time) on the other. The number of aircraft/flights served under such conditions represents the airport's "practical" capacity.

Externalities

In general, congestion and delays can be expressed as externalities by the cost of delays for airlines and their users—air passengers. In such contexts, usually only delays over 15 min are deemed relevant. Their costs usually depend on the airline/aircraft operating costs and the value of passengers' time. For example, the unit cost of marginal delays implying both primary and reactionary delays estimated depending on aircraft size are as follows: $C_a(S) = 0.10S - 0.167$ (€/min), ($R^2 = 0.92$), where S is the number of aircraft seats ($40 < S < 450$). The average value of time for an average passenger using all categories of aircraft is estimated to be $\alpha =$ €39/min/pass [17, 28].

Mitigating Measures

Congestion and delays at a given airport can be mitigated through tactical and operational measures in the short-term and through strategic measures in the long term. The former include matching the time pattern of particular atms to the existing airport's runway capacity or number of slots in the most feasible way by using the Arrival and Departure Flow Management component of ATM (Air Traffic Management). This enables more efficient and effective utilisation of the available airport airside capacity. On the other hand, strategic measures generally imply increasing the airport's runway capacity. One approach encompasses introducing advanced technologies supporting more innovative operations, increasing runway capacity and mitigating airlines and air passenger congestion and delays without increasing airport size and consequent land take (see Chap. 5). Another approach implies physical expansion of the airport capacity by building additional airside and landside infrastructure such as runways, taxiways, aprons/gates, and/or passenger and cargo terminals. In addition, at airports developed into true multimodal transport nodes, relieving runway capacity pressure can be achieved by substituting some short-haul flights with equivalent, usually HS (High Speed) surface transport services (as described in Chap. 4).

3.2.4.4 Land Take (Use)

Source and Nature

The air transport system takes land for its infrastructure such as airports and ATC/ATM buildings, facilities and equipment. Airports take most land in the scale of hundreds and thousands of hectares. Figure 3.2 shows a simplified scheme. As can be seen, the size of land taken depends mainly on the number and length of runways and their spatial configuration. In general, six typical (theoretical) generic airport configurations exist at airports worldwide, namely: single runway; two parallel runways both used for landings and take-offs; two parallel runways of which one is used for landings and another for take-offs; two converging runways each used for both landings and take-offs depending on the direction of the prevailing wind; two parallel plus one crossing runway each used for landings and take-offs; and two pairs of parallel runways of which two outer runways are used for landings and the two inner runways for take-offs. In addition, the particular symbols mean the following: A is the area of land taken (ha or km^2); d is the width of the runway strip (m); h is the width of the airport landside (terminal) area (m); l is the length of the airport landside (terminal) area (m); d_0 is the distance between the centrelines of two parallel runways; d_{01}, d_{02} is the distance between the centrelines of the first and second pair of inner and outer parallel runway(s), respectively; and α is the angle between a pair of converging/diverging runways. The minimal (standard) values of particular layout parameters for different airport categories are specified as recommendations. For example, these parameters for airports handling the largest aircraft (Category D and E) are as follows: $d = 300$ m, $h = 500$ m, $l = 500$ m; $L = 4{,}500$ m; $d_0 = 2{,}000$ m; $d_{01} = d_{02} = 1{,}050$. Consequently, the area of land taken can be computed as: $A = 260$ ha for configuration (a), 1,035 ha for configuration (b) and (c), 878 ha for configuration (d), 1,179 ha for configuration (e) and 1,980 ha for configuration (f). The differences obviously occur due to the number of runways.

As mentioned above, the volumes of existing and predicted airport traffic need to justify the area of land taken, which in turn will provide its appropriate utilisation. In addition, in most airports, the actual land taken is usually greater than the above-mentioned theoretical ideal cases. Illustrative examples are the recently built offshore airports in Japan—Kansai and Centrair—located on artificially created islands of 560 and 470 ha, respectively. In addition to the size of land taken, an important characteristic is the intensity of use of taken land, which is usually expressed as the volume of activities performed for accommodating traffic during a given period of time [10]. In general, given the fixed size of land occupied by a given airport, the intensity of land use increases in line with the volume of airport traffic. When the existing infrastructure capacity reaches saturation, the increments (runway, terminal building) are added. Such expansion causes a temporary drop in the intensity of land use before recovering with further traffic growth.

The intensity of land use at a given airport can also be expressed in terms of the volume of p-km (passenger kilometres) or t-km (ton-kilometres) realised per unit of land taken (ha). For such purposes, the structure of incoming and outgoing atms (flights) (short, medium, long haul) and the number of passengers (and the amount of freight) on board need to be available.

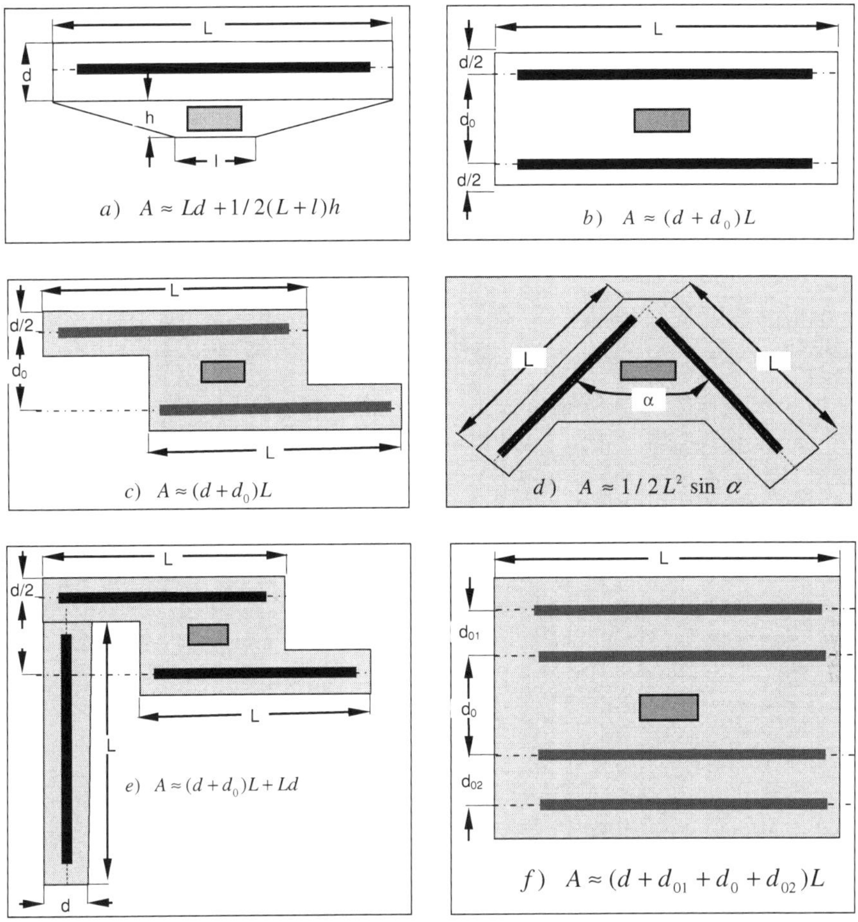

Fig. 3.2 Illustration of different airport configurations (layouts) (Compiled from [29])

Externalities

Existing and additional land taken support the growth of a given airport's traffic, generally causing both the airport's costs on the one hand and its revenues on the other to increase. Therefore, considering land taken (used) as a pure externality is ambiguous. The question is whether it is more socially acceptable to take land for an airport or to use it for some other economical or non-economical purposes such as for housing, agriculture, recreation, and the natural environment (green area with flora and fauna intact). In all these mutually exclusive cases, the land taken has a certain value, which may have economic, non-economic, and/or other market-based value. In particular, the economic value of land is relevant when assessing its social costs. For example, let: R_{ai} and C_{ai} be the total social revenues and costs, respectively, from operating a given airport occupying an area of land

S_i; R_{ji}, and C_{ji} be the social revenues and costs, respectively, from carrying out some other economic and/or non-economic activity (j) on the same land S_i. The value (i.e. cost) of one unit of the airport land can be determined as follows [28]:

$$C_l = \frac{\left[(R_{ai} - R_{ji}) - (C_{ai} - C_j i)\right]}{S_i r} \tag{3.5}$$

where

r is the capitalisation rate converting future monetary values into present value.

The nominator of expression (3.4) is often called the annual return of land. In addition, expression (3.4) reflects the intensity of land use in monetary terms.

Mitigating Measures

The most effective mitigating measure for land use (take) by a given airport is its full incorporation in regional medium to long term plans as well as their strict adherence. This enables reserving sufficient land for the airport's eventual expansion. In addition to this, "buffer" zones free of housing also need to be planed, which can in turn additionally contribute to mitigating the impact of airport noise on the surrounding population.

3.2.4.5 Waste

Source and Nature of Impact

At airports, waste is generated in quantities usually positively correlated to traffic volumes. Airport waste can be broadly classified as solid or liquid waste. In addition, it can be further separated into non-industrial and industrial waste. The former originates from passenger catering services provided on-board the aircraft, and from consumption by airport employees and visitors (food, newspapers, cans, paper). The latter originates from daily activities such as washing and cleaning aircraft and other ground vehicles, aircraft and engine maintenance, repair and testing including painting and metal work, aircraft de-icing, and maintenance operations on ground vehicles. This waste is further categorised into hazardous and non-hazardous waste. The former is managed according to the strict national and airport regulations governing collection, treatment, storage, and disposal [18].

Externalities

Airport waste can be considered an externality when it causes further damage to people's health and the environment. In particular, incidental leakage of hazardous liquid industrial waste such as aviation fuel, oil, and vehicle washing and cleaning liquids can contaminate the soil and drinking water, and consequently endanger the health and even the lives of people and natural habitats. In such cases, externalities are counted as the costs of eliminating such damage in the broadest sense. This usually implies cleaning up the contaminated areas and eventually strengthening protective infrastructure and preventive measures.

Waste mitigating measures are a part of the airport waste management system, which exists at almost all airports. The airport waste management system is usually designed and operated in accordance with applicable national and local legislation. This particularly refers to storing and disposing of waste in dedicated areas which cannot be used for other, more profitable activities [18]. The system usually includes identification of sources, location, types and quantity of waste generated, the infrastructure, facilities and equipment to deal with different types and quantities of waste, and finally the efficiency and effectiveness of waste collection, storage, recycling, and disposal.

An efficient and effective waste management system usually implies waste avoidance, minimisation, and recycling. This involves sorting waste at the collecting locations, i.e. at source, into solid and liquid, hazardous, and non-hazardous waste, reducing generated quantities, continuous increase in reuse, recycling and reprocessing of waste materials, and permanent improvement of waste management practices. In particular, recycling implies conversion of waste into energy through thermal treatment (processing). For example, at Frankfurt Main airport (Germany), the usage rate of recycled waste has increased from 50% in the year 1995 to about 85% in the year 2005. Consequently, as one of the final objectives, the cost of waste management was reduced.

3.3 Indicator System as the Core of the Methodology

3.3.1 Background

The indicator system as the core of the methodology for monitoring, analysing, and assessing the intensity of greening, i.e. sustainability of airports, is inherently complex due to the following factors:

- Performance multidimensionality, which implies considering the given airport as a system with numerous (inherently) interdependent and diverse components, involved parties, effects (benefits), and impacts (costs);
- Complexity in setting greening up, i.e. determining sustainability targets due to the above-mentioned interdependency;
- Complexity in assessing the marginal and global contribution of particular policy measures, and advanced technologies and operations on the intensity of greening, i.e. more sustainable development in the medium to long term.

Therefore, an effective and efficient indicator system consists of indicators and their measures, which should satisfy the following criteria: [28]

- Sufficient generosity in order to be applicable to airports at different geographical (national and international) locations and involved parties;
- Based on available data for their quantification;
- Clearly understandable for particular involved parties and able to be calculated in a relatively short period;
- User-driven, i.e. provided for the intended users and/or audience, and policy-relevant, i.e. pertinent to policy concerns; and
- Sufficiently aggregated, meaning that there should be few final indices.

This indicator system is also based on the following assumptions:

- The system should reflect the preferences of particular parties involved in dealing with greening, i.e. the sustainable development of a given airport;
- The system consists of four sub-systems corresponding to the four different dimensions of the system's performance (operational, economic, environmental, and social);
- The particular indicator measures express effects (benefits) and impacts (costs) of operations of a given airport quantitatively in either absolute or relative monetary or non-monetary terms, usually as functions of the volumes of the airport's traffic accommodated during a given period;
- If "targets" are determined for particular indicators and measures, they will be used as benchmarks for assessing the current level of greening, i.e. sustainable development; and
- The indicators and their measures can be inherently dependent on each other as well as on particular influencing factors.

3.3.2 Prior Research

Contrary to specific research conducted by international aviation organisations and consultancy bodies, academic research on developing a framework and an indicator system as the core of the methodology for systematic monitoring, analysing, and assessing greening, i.e. sustainable development of airports, has been relatively scarce to date [32]. In the given context, scarcity particularly relates to development of a system of indicators and their measures for quantifying the particular above-mentioned effects (benefits) and impacts (costs), as well as for predicting their prospective balance, depending on related influencing factors. Nevertheless, some efforts are noteworthy, e.g. the development of indicators for planning sustainable development of the transport system [31]. In addition, research had been carried out in elaborating the concept of sustainable aviation and its development [35]. Furthermore, research on developing a methodology for assessing the sustainability of the air transport system consisting of airports, air traffic control, and airlines, has been carried out [26]. Methodology has been set up at the conceptual level, its core represented by an indicator system for quantifying the performance of particular components of the air transport system. The infrastructural, technical, operational, economic, environmental, social, and institutional/policy performance were considered as relevant respecting the attitudes of the involved parties. This methodology, consisting of about 104 indicators and its measures, was soon partially applied to different components of the air transport system in order to illustrate its potential and feasibility for eventual practical planning applications [27, 28]. Subsequent additional research resulted in developing indicators of sustainable development of civil aviation. In this research, 29 appropriately modified indicators were derived from other transport modes and defined respecting the interests and attitudes of particular parties involved in the UK [20]. Most recently, research focusing exclusively on developing an indicator system for monitoring, analysing, and assessing greening, i.e. more sustainable development of airports, has been carried out [30]. This chapter contains a substantive part of this latest research.

3.3.3 Scope and Structure

The indicator system as the core of the methodology for monitoring, analysing, and assessing the level of greening of airports consists of indicators and measures reflecting their operational, economic, environmental, and social performance. In the given context, indicators of operational, economic, and social performance are additionally considered due to their interdependent relationship and influence on environmental factors. Accordingly, the process of greening, i.e. sustainable development of airports, becomes more consistent and transparent. Indicators and measures for infrastructural and technical and technological performance are not particularly considered. These could be easily developed, for example, by considering the level and quality of information, facilities, and equipment for servicing passengers and their baggage in the terminal buildings, the rate of mechanisation of the air cargo handling and storage processes, utilising natural gas, LH_2 (Liquid Hydrogen), and/or electrical-powered vehicles in servicing passengers and aircraft in the airport airside area, and passengers and freight in the airport landside area (ground access systems).

The indicators and measures within the scope of the methodology can be estimated for a given airport and/or for the several airports serving a given region of the airport system [28].

3.3.3.1 Operational Indicators

The indicators of operational performance include "demand", "capacity", "quality of service", and "integrated multimodal services".

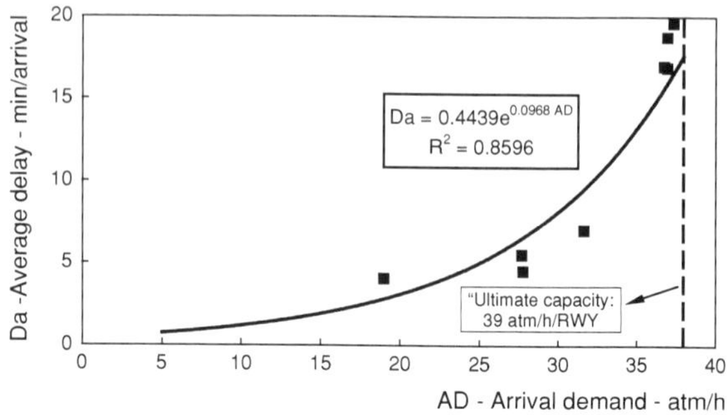

Fig. 3.3 Relationship between demand, capacity, and average flight delay at London Heathrow Airport (UK) (period 2000–2007) (Compiled from [3, 29])

- *Demand* indicates the scale of airport operations. The number of air transport movements (atm), passengers, and the volume of freight shipments accommodated during a given period of time (hour, day, year) can be the measures of this indicator. Sometimes, it is more convenient to use the above-mentioned WLU as an aggregate measure. The airport operator prefers these measures to be as high as possible and to increase over time [10].
- *Capacity* reflects the airport's capability to accommodate a certain volume of demand under given conditions. Two measures can be used: the airside capacity in terms of the maximum number of atm, and the landside capacity as the maximum number of WLUs accommodated over a given period of time (hour, day, or year). Both can be expressed as either "ultimate" or "practical" capacity. The former implies conditions of constant demand for services while the latter implies conditions of imposing an average delay on each unit of demand. It is preferable that both be as high as possible and increase in line with the growing demand.
- *Quality of service* reflects the relationship between the airport demand and "practical" capacity. Generally, the average delay per atm or WLU, which occurs whenever demand exceeds capacity, can be used as a measure which is preferred to be as low as possible and to decrease as demand increases. Figure 3.3 shows the development of the relationship between demand and capacity in terms of atm at the large and congested European airport London Heathrow UK [29].
- As can be seen, the average delay per atm increases at an increasing rate with demand. The intensity of accommodated demand at which the average delay is guaranteed to each aircraft/flight represents the airport's "practical" or "declared" capacity, which in this example is 78 atm/h (i.e. 39 atm/h/runway). This system of two parallel runways operates in "segregated" mode (i.e. one runway is used exclusively for arrivals and the other exclusively for departures). Currently, the average delay per atm during the 10 busiest hours of the day is about 18 min [3]. In particular, the above-mentioned development has proved the theory of "practical" airport runway capacity by showing a very similar, if not identical, type of the delay–demand-capacity relationship [9];

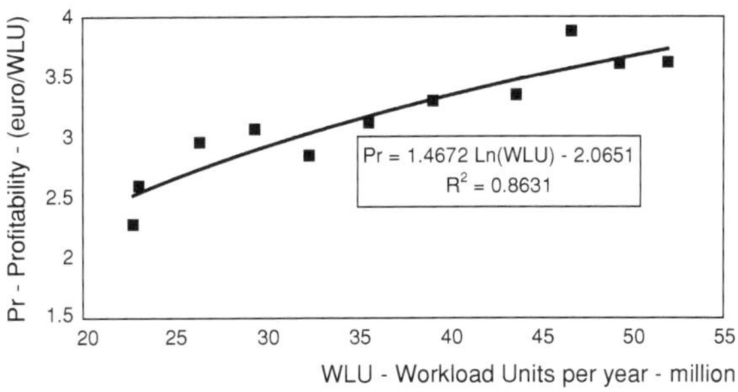

Fig. 3.4 Profitability vs. Traffic volume of Amsterdam Schiphol Airport for the period 1990–2000 (Compiled from [34])

- *Integrated multimodal services* are an indicator which may be relevant for airports connected to surface regional, national, and international transport networks. Generally, these airports have the opportunity to reduce congestion and aircraft and passenger delays, as well as improve capacity utilisation, by substituting some short-haul flights with adequate surface, usually conventional and/or HSR (High Speed Rail) services on the one hand, and using the freed slots for more profitable long-haul air services on the other. For example, three European hubs–Frankfurt Main, Paris CDG and Amsterdam Schiphol airports—are connected to the Trans-European HSR network enabling the above-mentioned air/rail substitution as described in Chap. 4 [11, 21, 25, 30].

3.3.3.2 Economic Indicators

As a business enterprise, any given airport must ensure favourable economic performance, of which "profitability" and "labour productivity" are the most convenient indicators:

Profitability reflects the airport's financial and operational success. It is usually measured in terms of operating profits, i.e. the difference between operating revenues and costs per unit of airport output—WLU [10]. It is preferred when this figure is as high as possible and when it increases proportionately to the airport's output. Figure 3.4 shows an example of profitability of Amsterdam Schiphol airport (The Netherlands). As can be seen, the airport's profitability in terms of €/WLU has increased at a decreasing rate during the observed period, indicating the airport's diminishing long term levels of the annual marginal contribution.

Labour productivity reflects the efficiency of labour use at a given airport. The most convenient measure is the number of WLUs or ATMs per direct airport employee carried out over a given period of time (year) [10, 28]. This measure is preferred to be as high as possible and to increase together with the number of employees. Figure 3.5 shows an example for Amsterdam Schiphol airport (The Netherlands). As can be seen, during the observed period, labour

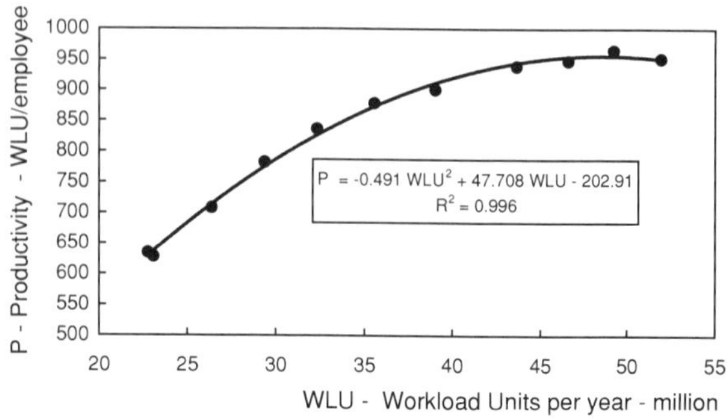

Fig. 3.5 Labour productivity vs. traffic volume at Amsterdam Schiphol Airport for the period 1990–2000 (Compiled from [34])

productivity has generally increased together with the number of WLU (albeit at a decreasing rate), and becomes zero after the annual number of WLU exceeds 45 million.

3.3.3.3 Social Indicators

Both direct and indirect employment at the given airport is considered an indicator of the social dimension of its performance. This indicator can be represented by the causal relationship between the total number of employees at and around the given airport and the annual volume of airport traffic. Some examples of the relationship for both direct and indirect employment across selected European airports are as follows [1, 28]:

1. Direct employment

$$E_d(q) = 1.4702q - 4.209; \quad R^2 = 0.901; \quad N = 22 \tag{3.6a}$$

2. Total employment

$$E_t(q) = 0.577q^{1.493}; \quad R^2 = 0.930; \quad N = 22 \tag{3.6b}$$

where

q is the annual number of passengers handled at a given airport (million); and
N is the number of elements in the sample.

Expression (3.6a, 3.6b) indicates that the number of staff directly employed at any given European airport amounts to about fourteen hundred employees per million passengers, which is about 40% higher than the commonly used ratio of around ten hundred per million. In addition, the total number of employees has increased disproportionally with the volume of airport traffic.

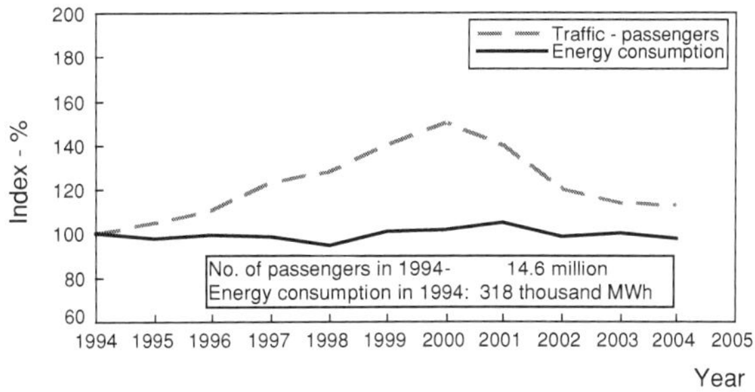

Fig. 3.6 Energy consumption and air passenger traffic at Zurich airport (Switzerland) (Compiled from [36])

3.3.3.4 Environmental Indicators

Energy consumption, greenhouse gas emissions (air pollution), noise, land use, and waste efficiency are all considered as indicators of the environmental dimension of airport performance:

- *Energy efficiency* relates to the total energy consumed by a given airport over a given period of time (year). In case of airport airside activities, this includes energy consumed by the aircraft LTO cycle, as well as by the ground vehicles and equipment servicing the aircraft at the apron/gate complex. In case of landside activities, this includes energy obtained from different sources, for lighting and heating of the terminal and other buildings supporting operations. In the wider sense this can also include energy consumption by the airport ground access systems/modes (see Chap. 7). A useful measure for this indicator in any of the above-mentioned cases can be energy consumed per unit of airport output, i.e. atm, passenger, cargo unit, and/or WLU, accommodated over a given period of time (year). This measure is preferred to be as low as possible and to decrease with the volume of airport output. Figure 3.6 shows an example of managing electricity consumption in the passenger terminal buildings at Zurich airport (Switzerland), where the airport aims to maintain electricity consumption at 1994 levels. As can be seen, this objective has been achieved despite an increase in the number of air passengers of about 10–20%, and the building surface area of about 40% [36]. Consequently, the energy consumption per passenger has decreased during the observed period.
- *Emissions, i.e. air pollution efficiency* relates to the total emissions of greenhouse gases generated by the operation of a given airport. As in the case of energy consumption, the amount of all or only certain specific emissions of air pollutants in both airport airside and landside areas from the above-mentioned energy consuming sources and activities can be taken into account [17]. For example, the quantity of emissions per polluting event in the airport airside area, i.e. LTO cycle, can be used as the standard measure recommended by the ICAO [23]. The non-LTO cycle-related pollution could be allocated to each of them. Figure 3.7 shows energy consumption and related emissions of CO_{2e} (Carbon Dioxide equivalents) for different aircraft types during the LTO cycle.

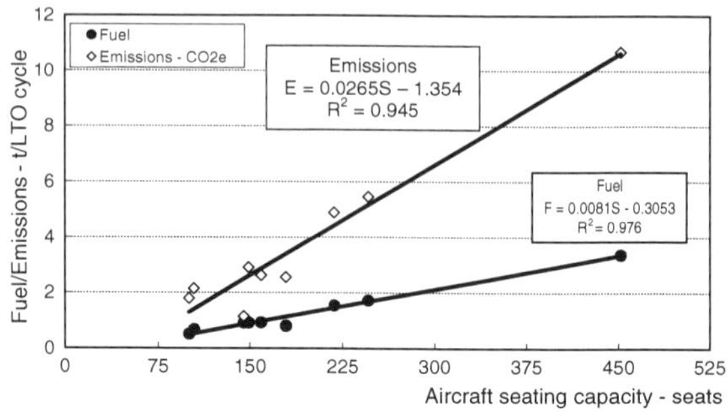

Fig. 3.7 Fuel consumption and CO_{2e} emissions during the LTO cycle depending on aircraft size (Compiled from [28])

Table 3.1 An estimation of the area, population, and number of households within the noise contours L_{den} at London Heathrow Airport

Contour level dB(A)	Area (km^2)	Population (000)	Household (000)
>55	302.3	782.9	344.9
>60	114.3	260.5	109.8
>65	47.7	74.5	29.9
>70	20.8	16.6	6.5
>75	7.5	1.7	0.7

Compiled from [7, 28]

The standard time components of the LTO cycle are: 0.7 min for take-off, 2.2 min for climb, 4.0 min, for approach, and 26.0 min for taxi/idle phase of operations [28].

As can be seen, both factors increase in line with aircraft size (seating capacity). Consequently, airports handling greater number of LTO cycles carried out by larger aircraft are faced with higher energy consumption and related emissions of greenhouse gases in their airside activities. Nevertheless, this measure is preferred to be as low as possible and to decrease with the introduction of the above-mentioned mitigating measures. On a wider scale, total emissions of greenhouse gases, generated by all incoming and outgoing aircraft/flights connecting a given airport to the rest of the air transport network during a given period of time (one year), can be used as an additional measure.

- *Noise Efficiency* relates to the noise energy generated by the atms (air transport movements) and related aircraft-servicing operations at the apron/gate complex in the airport airside area and the noise energy generated by ground access systems/modes in the airport landside area (see Chap. 7) during a given period of time. Some of the measures for this indicator are the size of the affected area (km^2), population, and the number of households exposed to the equivalent long term noise level L_{eq} (dB(A)–decibels). This indicator is preferred to be as low as possible and to diminish with the number of atms. Table 3.1 gives an example of reducing noise, the exposed area, related population, and the number of exposed households around London Heathrow Airport (UK).
- *Land use efficiency* relates to utilisation of land acquired for building and operating a given airport–both airside and landside areas. Once the

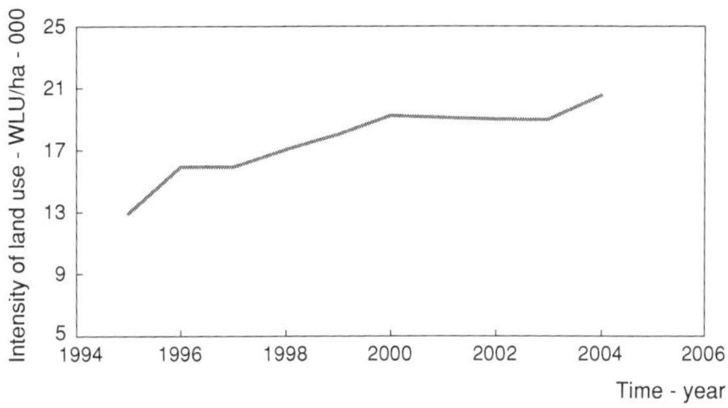

Fig. 3.8 Land use efficiency at Amsterdam Schiphol Airport (Compiled from [28, 34])

infrastructure has been constructed, the intensity of use of acquired land becomes dependent on the volume of accommodated traffic. However, this intensity is always limited by the infrastructure capacity. In such a context, a convenient measure for this indicator is the volume of WLUs accommodated during a given period of time (year) per unit of acquired land. This measure is preferred to be as high as possible and to increase with the area of land occupied by the airport. Figure 3.8 shows an example of the land use efficiency at Amsterdam Schiphol airport (The Netherlands) before and after building the new (fifth) runway in 2002.

As can be seen, before 2001 (the year of the aviation crisis caused by the September 11 terrorist attacks on the US), the intensity of land use increased due to increased air traffic volumes. Over the next three years, intensity of land use stagnated due to a combination of factors including stagnation of traffic growth and opening of the new (fifth) runway (2002). The addition of the new airside infrastructure actually increased the area of land used by the airport. Nevertheless, the intensity of land use later recovered due to recovering and continuing air traffic growth.

- *Waste efficiency* relates to the waste generated by the given airport's operation. This can include or exclude airline-generated waste [5]. In any case, a convenient measure of this indicator is the total waste generated by the airport during a given period of time and/or the quantity of waste generated per unit of the airport's output (passenger/and/or WLU). These measures are preferred to be as low as possible and to decrease with the airport's output over a given period of time (year). Figure 3.9a, b shows an example of the waste (in)-efficiency across seven UK airports operated by BAA (British Airport Authority) (UK). Specifically, Fig. 3.9a shows the dependence of the total annual quantity of generated waste on the annual number of accommodated passengers.

  As can be seen, these total quantities have increased almost linearly with the number of air passengers, indicating that larger airports usually generate larger quantities of waste, and vice versa. Figure 3.9b shows that the average quantity of waste per passenger across the same airports has also increased disproportionally with the annual number of passengers. This development has occurred mainly due to an increasing proportion of long-haul flights at larger airports, which

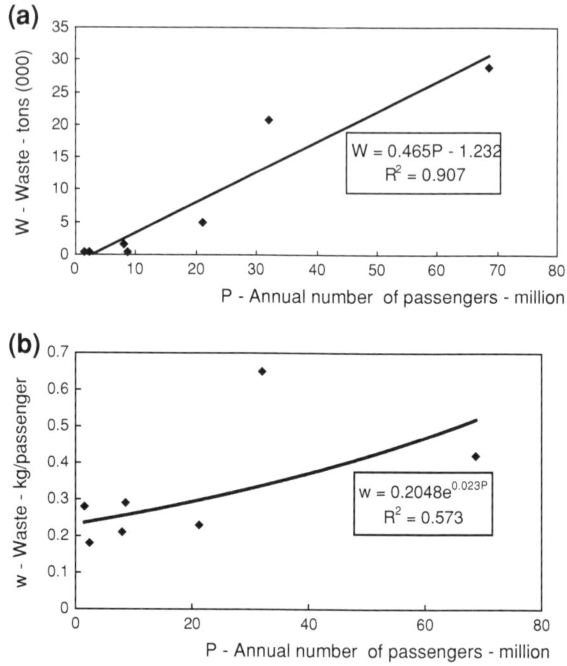

Fig. 3.9 Waste efficiency (selected UK airports in 2005) (Compiled from [5]). **a** Total quantity of waste. **b** Relative quantity of waste

demand increased in-cabin service per passenger. In the given case, about 23% of the total quantity of waste is recycled [5].

3.3.4 Some Applications of the Indicator System

The above-mentioned indicator system has been applied to indicate the potential trade-off between particular effects (benefits) and impacts (costs) across different airports. As such, this trade-off could be of relevance for policy makers, airport operators, and authorities, who determine the current and prospective airport development. Application is illustrated by two examples. The first example represents an analysis of the prospective relationship between airport operational capacity (as an indicator of operational performance) and the noise and/or air pollution cap (quota) set to constrain the environmental and social impacts (as indicators of environmental performance). For this purpose, in addition to the above-mentioned concepts of "ultimate" and "practical" capacity, the concept of "environmental" capacity is defined as the maximum number of atms, passengers, and/or freight (i.e. WLUs) accommodated at a given airport during a given period of time under conditions of constant demand for services and within the specified environmental cap(s) (quota(s)). The second example describes some principles of trade-off between the total social benefits and costs while developing and operating a given airport on the one hand and simultaneously making it greener on the other.

3.3.4.1 Operational Capacity Versus the Noise Cap (Quota)

With respect to noise, the airport "environmental" capacity can be defined as the maximum number of atms accommodated during a given period of time under conditions of constant demand for services while generating the total sound energy within the prescribed limit, i.e. the noise cap (quota). The cap (quota) can be set differently for arrivals and departures. Thus, for those arrivals and departures carried out during period *T*, the "average energy sound level", or the "equivalent continuous noise" index $L_{a/eq/T}$ and $L_{d/eq/T}$, respectively, can be used. The L_{eq} index is designed to accumulate all aircraft sound energy for multiple noise events, either arrivals or departures realised during a given period of rime (1, 8, or 24 h). For example, in the UK, the concept of L_{eq} is applied to a 16 h period of daytime [3, 4, 8]. The cumulative sound energy contained in L_{eq} is assumed to be uniformly distributed over the time period *T*, and at most airports, it is different for daytime and night-time periods.[1] Analogously, $L_{a/*/T}$ and $L_{d/*/T}$ represent noise in *dB*(*A*) generated by an individual noise event, i.e. by an arriving and departing aircraft of type (*), respectively, during period *T* [15]. This noise is usually estimated at noise reference locations, which may be either the aircraft noise certification points or some other preselected locations in the vicinity of the given airport [23].

As noise quotas $L_{a/eq/T}$ and $L_{d/eq/T}$ are set according to the maximum level of tolerance of the affected population, the airport environmental capacity in terms of noise can be determined as $C_{a/T} = T \circ 10^{L_{a/eq/T}/10}$ for arrivals and $C_{d/T} = T \circ 10^{L_{d/eq/T}/10}$ for departures. The period *T* is expressed in seconds. Similarly, the sound energy of an individual noise event—arrival and/or departure—can be expressed as $N_{i/a} = 10^{L_{a/k/T}/10}$and $N_{i/d} = 10^{L_{d/k/T}/10}$, respectively. By dividing the airport noise capacity with the average sound energy per individual event, the number of aircraft movements satisfying the prescribed noise quota during period *T* can be estimated as: $C_{e/n} = C_{a/T}/N_{i/a-d}$. Figure 3.10 shows an example of the potential relationships [28].

As can be seen, for a given noise quota, the airport's noise capacity in terms of the number of arrivals decreases with the average noise per individual arrival. In addition, by increasing the available noise quota, the airport noise capacity increases given the average noise per individual event. The average noise per individual event depends on the structure of the aircraft using the airport as well as on the proximity of the flight path to noise measurement locations. Generally, the average noise per event will be higher with a greater proportion of heavy and Category 2 noisier aircraft in the arrival/departure mix of aircraft and when the noise measurement locations are closer to the flight path [23].

Under specific circumstances, for example in case of severe night limitations or a complete night-flight ban, the noise cap (quota) may also act as a true constraint

---

[1] At airports where a night flight ban is imposed, the noise quota is zero during the ban period. During the day, it is above zero, e.g. 57 dB(A) at London Heathrow, 85 dB(A) at Birmingham, and 73 dB(A) at Frankfurt airport [8].

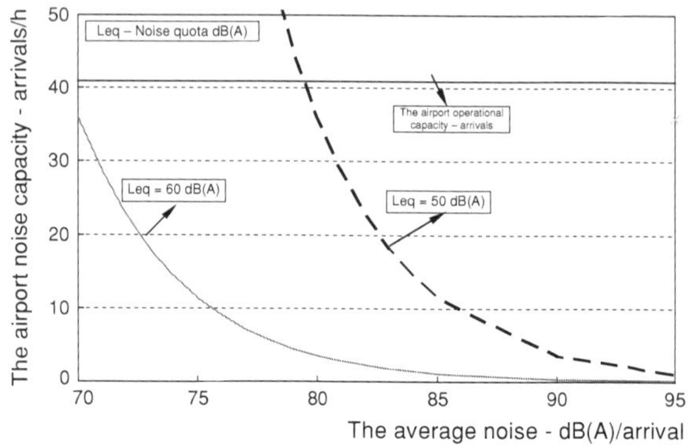

Fig. 3.10 Airport noise capacity depending on the average noise per arrival and the noise cap (quota) (Compiled from [30])

to the airport's capacity. The dotted line in Fig. 3.10 illustrates that in such cases, if the noise quota is set at the level of minimal exposure (for example 50 dB(A)), only a few—if any—aircraft will be given access [4]. In contrast, if the average noise per event is set at 80 dB(A), the number of arrivals within the given noise quota of 60 dB(A) will be about 40 ops/h. Generally, relaxing the noise quota increases the number of runway operations at a given airport up to its full operational "ultimate" and/or "practical" capacity, which is in the given case 41 arr/h.

3.3.4.2 Operational Capacity Versus the Air Pollution Cap (Quota)

The airport's environmental capacity with respect to a given air pollution cap (quota) can be expressed as the maximum number of atms and/or WLUs achieved during a given period of time under conditions of constant demand for service, providing total air pollution remains within the cap. This implies that if this cap during period T is $Q_{m/T}$ and the total air pollution increases proportionally to the traffic volume, the capacity achieved within the prescribed cap can be estimated as: $C_{e/ap} \approx Q_{m/T}\, n_a\, e_a\, q_{ap}$, where: n_a is the average number of air pollution activities per unit of traffic (atms, WLUs); q_{ap} is the average energy consumption per activity; and e_a is air pollution per unit of consumed energy. In addition, this capacity can be considered in a more complex form. For example, particular activities, energy consumption, and related air pollution can be separately analysed (quantified) for the airport airside and landside areas. In the airside area, this capacity can be expressed by the "number of atms/T" or by the "number of LTO cycles/T".[2] In the landside area, this capacity can again be expressed by the "volume of WLU/T".

[2] Emissions include greenhouse gases generated in the airport airside area during aircraft LTO cycles and by aircraft ground servicing vehicles [7, 13, 16, 23].

3.3.4.3 General Relationship Between Effects and Impacts

Airport growth usually brings increasing benefits to the airport operator, local community, and society. The size and structure of these benefits, mainly expressed in terms of local employment and consequent direct contribution to GDP, and by revenues gained from visitors to the region, are usually proportional to the volume of airport traffic accommodated under given conditions. This traffic, satisfied by airport operational capacity, may be faced with acceptable congestion and delays. Particular environmental constraints, such as noise and/or air pollution mentioned above may affect (limit) the airport operational capacity, which, under a given level of traffic (demand), may increase congestion and delays. If such constraints are in place for longer periods, such caps may limit the overall volumes of traffic, and consequently affect the airport's medium to long term growth, as well as expected effects (benefits). In addition, charging particular externalities in the form of taxes may increase airfares and consequently deter some passengers and airlines from using a given airport, which again in turn, may affect its growth. Any cap on using land for expansion of the airport airside and landside infrastructure directly affects the airport's operational capacity, congestion and delays, and consequently the airport's medium to long term growth, i.e. the related overall effects (benefits). Therefore, trading-off between particular effects–benefits and related impacts–costs by using different policy instruments (tools) need to be carefully carried out respecting their rather strong inherent mutual interrelations. In some sense, the above-mentioned system for monitoring, analysing, and assessment of greening, i.e. more sustainable development of a given airport, could be used for such a purpose when both total and/or partial effects (benefits) and impacts (externalities) are expressed in monetary terms. In such a case, the ratio $r = R/C$ can be used, where R is the total social benefits and C is the total social cost either from already realised or perceived (forecasted) operations of a given airport during a specified period of time (one or several years). Under such conditions, if $r > 1$, the airport will develop within the greening model, i.e. in a more sustainable way; if $r = 1$, airport development will be "neutral" or "zero" in terms of greening or sustainability; and finally if $r < 1$, airport development will not occur in the direction of greening, i.e. it will be unsustainable. This ratio could also be used to evaluate the investment feasibility for any given airport.

3.4 Concluding Remarks

This chapter has described the methodology for monitoring, analysing, and assessing the process of greening, i.e. sustainable development of airports. This methodology has been based on the concept, strategies, and tactics of greening, as well as an indicator system for quantifying the main effects (benefits) and impacts (costs). The concept, strategies, and tactics of greening have dealt with identifying particular effects–benefits and impacts–costs (externalities) created by airport operations, explaining their importance and main influencing factors including their mutual relationship (dependability), and their balancing over the medium to long term. The indicator system consists of indicators and their measures reflecting the airport's operational, economic, social, and environmental performance. Quantification of some of these indicators and measures has indicated that in the given cases, airports have become increasingly greener over time by generally increasing the effects (benefits) and diminishing some impacts (costs) mainly in relative terms, i.e. as quantity per unit of output. This general observation however does not apply to congestion and delays, and waste.

In some cases, trade-offs between particular effects–benefits and impacts–costs (externalities) have indicated their high mutual dependency. For example, strict caps on impacts of noise, greenhouse gas emissions, and land take (use) for airport infrastructure expansion might ultimately constrain the volume of airport operations in the short- (daily) but also in the medium to long term period (years), thus affecting airport growth and limiting the related effects–benefits.

Existing and future policies, strategies and tactics of greening, i.e. sustainable development of airports, can consider the proposed methodology as a component of a more sophisticated "tool" for managing the greening process. The output of this "tool" should enable particular parties involved such as airport operators, local and national authorities and communities, airlines, and users—air passengers and freight shippers—to assess their individual contribution and eventual future role in the greening processes.

In addition, specific advanced airport developments are expected to further consolidate their greening processes. Specifically, in the airport airside area, these include: introducing advanced procedures for increasing airport airside (runway) capacity with a view of reducing airside congestion and airline and air passenger delays, thus temporarily mitigating the need for building additional runways requiring additional land take, handling more fuel efficient aircraft, improving aircraft guidance during LTO cycles in terms of both time and corresponding fuel consumption, replacing APUs with electric units which will allow aircraft to switch off their engines, and lobbying for introducing alternative fuels in the long term period. In the airport landside area, such developments include: consolidating and promoting public mainly light- and heavy-rail transport systems/modes for airport short- medium, and long-distance accessibility, and increased use of the wind and solar energy for heating, cooling, and lighting buildings, which is expected to mitigate the energy consumption and related emissions of greenhouse gases. Finally, airports need to be included in various schemes of charging for greenhouse gas emissions, as this will stimulate airports on the one hand and force them to firmly remain on the greening, i.e. sustainable development medium to long term trajectory on the other. Some of the above-mentioned developments are elaborated in the forthcoming chapters.

References

1. ACI Europe (1998) Creating employment and prosperity in Europe. A study of the social and economic impacts of airports, Airport council international Europe, Brussels, Belgium
2. ACI (2008) Airports—actions on climate change. Airport council international, Montreal, Canada
3. ACL (2007) Peak periods at heathrow airport. Report submitted to CAA, attachment 1, Airport Co-ordination Ltd., London
4. Ashford N, Wright PH (1992) Airport engineering. Wiley, New York
5. BAA (2006) Generation I, environmental statement–waste. British Airport Authority, London, UK
6. Betancor O, Carlos Martin J (2005) Social noise cost at madrid barajas airport. 8th NECTAR conference, June 2–4, Las Palmas, Gran Canaria, Spain
7. CAA (2004) Noise mapping–aircraft traffic noise research study on aircraft noise mapping at heathrow airport conducted on behalf of defra. ERCD Report 0306, Civil Aviation Authority, London
8. DETR (2000) The future of aviation. The government's consultation document on air transport policy, Department of the environment, transport and the regions, London, UK
9. De Neufville R, Odoni A (2005) Airport systems planning, design, and management. McGraw Hill Book Company, New York, USA
10. Doganis R (1992) The airport business. Routledge, London, UK
11. EC (1998) Interaction between high-speed rail and air passenger transport. Final report on the action COST 318, EUR 18165, European Commission, Luxembourg
12. EC (1999) Air transport and the environment: towards meeting the challenges of sustainable development. European Commission, The economic and social committee and the committee of the regions, the European parliament, (COM), Brussels, Belgium
13. EC (2003) Guidelines on the revised interim computation methods for industrial noise aircraft noise, road traffic noise, railway noise, and related emission data, annex. The official journal of the European Union, Brussels Belgium, pp 50–64
14. EC (2005) Optimum procedures and techniques for the improvement of approach and landing—OPTIMA. Beliverable D1.2, 6th framework programme, European Commission, Brussels, Belgium
15. ECAC (1997) Report on standard method of computing noise contours around civil airports. European civil conference, ECAC. CEAC Doc 29, Sec Ed., Neuilly-sur-Seine Cedex, France
16. EEC (2005) Standard inputs for eurocontrol cost-benefit analysis. Eurocontrol, Brussels, Belgium
17. EPA (1999) Evaluation of air pollution emissions from subsonic commercial jet aircraft, EPA–420R–99–013, ICF consulting group, environmental protection agency, Washington, DC, USA
18. FAA (2001) Management of airport industrial waste. Federal aviation administration, U.S. department of transport, report no. 150/5320-15, Washington DC, USA
19. Fraport (2009) Frankfurt airport: climate protection and energy. Fraport AG, Frankfurt, Germany
20. Grimley PM (2006) Indicators of sustainable development in civil aviation. PhD Thesis, Loughborough University of Technology, Loughborough, UK
21. HA (1999) Environmental impact of air transport in comparison with railways and transport by rail in the European Union. Intermediate report, Hera—Amalthea, S.L., Madrid, Spain
22. ICAO (1993a) Aircraft noise. Environmental protection, annex 16, Vol 1. International civil aviation organisation, Montreal, Canada
23. ICAO (1993b) Aircraft engine emissions. Environmental protection, annex 16, Vol 2. International civil aviation organisation, Montreal, Canada
24. IPCC (1999) Aviation and the global atmosphere. Intergovernmental panel of climate change. Cambridge University Press, Cambridge, UK

25. INFRAS (2005) Sustainable avation, study carried out for air transport action group (ATAG). Juris Druck and Berlag, Zurich, Switzerland
26. Janic M (2002) Methodology for assessing sustainability of air transport system. J Air Transport 7(2):115–162
27. Janic M (2004) An application of the methodology for assessment of the sustainability of air transport system. J Air Transport 9:40–82
28. Janic M (2007) The sustainability of air transportation: a quantitative analysis and assessment. Ashgate Ltd Limited, Aldershot, UK
29. Janic M (2009) The Airport Analysis, Planning, and Design: Demand, Capacity, and Congestion. Nova Science Publishers, Inc. New York, USA
30. Janic M (2010) Developing a an indicator system for monitoring, analyzing, and assessing airport sustainability. Eur J Transport Infrastructure Res 10(5):205–225
31. Johnston R (2008) Indicators for sustainable transport planning. Transport Res Record (TRR) 2067:146–154
32. Longhurst J, Gibbs DC, Raper DW, Conlan DE (1996) Towards sustainable airport development. Environmentalist 16(3):197–202
33. Morrell P, Lu CH-Y (2000) Aircraft noise social cost and charge mechanisms–a case study of amsterdam airport schiphol. Transport Res D 5:305–320
34. Schiphol Airport (2004) Annual statistical review 1990–2004. ZG Schiphol, The Netherlands
35. Upham P, Maughan J, Raper D, Thomas C (2000) Towards sustainable aviation. EARTHSCAN, UK
36. ZA (2005) Energy management: zurich airport, unique -environmental services Zurich airport. Juris Druck and Berlag, Zurich, p 25

Chapter 4
Greening Airports II: Transforming an Airport into a True Multimodal Transport Node

4.1 Introduction

At many large airports, traffic growth has caused an increase of airside and landside congestion and airline and air passenger delays, noise, local emissions of greenhouse gases, and waste. Consequently, transforming such airports into true multimodal transport nodes is considered one of the long term alternatives for mitigating some of these impacts. What does this imply? Airports are, by definition, multimodal transport nodes, which enable air passengers and air cargo to transfer from the airports' ground access systems/modes to the air transport system/mode, and vice versa, as will be described in more detail in Chap. 7. Under such circumstances, developing such airports into true multimodal transport nodes implies connecting them to ground transport systems/modes, providing similar services to/from them as the air transport system/mode over the specified short- to medium-haul inter-city and inter-airport travel distances (routes). In this respect, air passengers can be transferred at the airport on two levels: (a) firstly, between the airport ground access systems and one of the longer distance transport systems/modes; and (b) secondly, between two inter-city and inter-airport transport systems/modes. By simultaneously providing transport capacities through at least two either competing or complementing modes APT (Air Passenger Transport) and HSR (High-Speed Rail), the airports in question could be considered to operate as true multimodal transport nodes.

In Europe, connecting airports to inter-city ground transport systems/modes has usually implied their inclusion into the regional, national, and international (conventional and/or HSR) transport networks. Some developments so far have implied connecting four European hubs—Frankfurt Main, Paris CDG (Charles de Gaulle), Madrid Barajas, and Amsterdam Schiphol airport—to the Trans-European HSR network, which has allowed them to operate as true multimodal transport nodes. There, although still on a relatively modest scale, substitution of some APT short-haul flights with equivalent HSR services either through modal competition or complementarity has already developed [12, 18, 26, 37, 40]. In general,

M. Janić, *Greening Airports*, Green Energy and Technology,
DOI: 10.1007/978-0-85729-658-0_4,

such substitution has removed particular APT short-haul flights from the airline and airport schedules and consequently reduced the overall airside congestion and delays and related costs for both airlines and air passengers. In addition, noise affecting local populations, energy consumption and related emissions of greenhouse gases, have been eliminated from these flights. Provided these impacts from substitution of HSR services are lower, net savings can be achieved. In addition, further growth of APT short-haul demand will likely be accommodated by HSR services instead of APT short-haul flights, thus consequently mitigating the medium to long term pressure for building additional airport airside infrastructure (runways) taking additional (relatively substantive) land.

The eventual mitigating of some of the above-mentioned social and environmental impacts and their costs through the developments described above would contribute to greening, i.e. sustainable development, of these airports in the medium to long term. However, the scale of any potential effects still needs to be assessed in further research, which is best performed on an airport-by-airport basis.

This chapter elaborates on the potential effects in terms of savings in airline and air passenger congestion and delays, noise, and emissions of greenhouse gasses, and their costs, which could be achieved by substituting some short-haul flights with equivalent HSR services at a large congested European airport assumed to operate under given "what-if" scenarios as a true multimodal transport node. These savings can be considered as benefits in evaluating the overall social-economic feasibility of developing a given airport into a true multimodal transport node.

4.2 Characteristics of HSR and APT in Europe

4.2.1 Development of HS Transport Systems

Compared to other transport modes, the main characteristic of HS (High-Speed) transport systems such as HSE and APT is their speed. In general, it can be said that progressively improving GDP and PCI (Gross Domestic Product and Per Capita Income respectively), as well as the forces of globalising the national and international economies, have driven the demand for increasing transport speed [29]. In particular, high-speed travel provides overall benefits to its users (passengers) in terms of faster and deeper market penetration, substitution of conventional transport modes, and improved use of time and monetary budgets.

APT emerged as the earlier of the two HS travel systems considered, as early as the 1950s. Its speed also increased continually until 1975 when the Concorde aircraft was launched, as shown in Fig. 2.1 (Chap. 2). In addition, the system has been permanently modernised and modified through aircraft capabilities, airline operational strategies, and governmental regulation supported by building new airports and modernising ATC/ATM systems, as described in Chap. 2 [29].

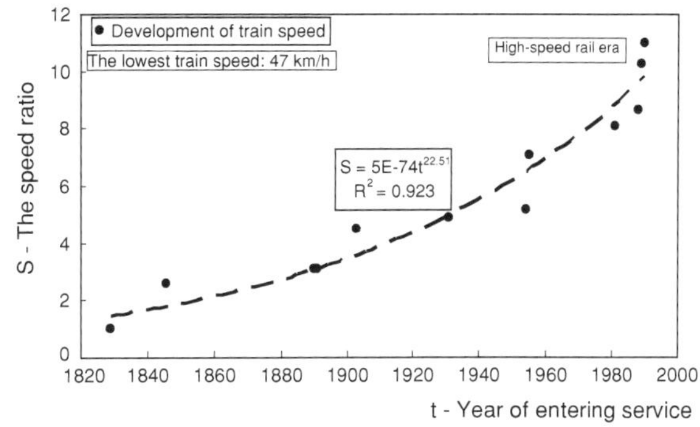

Fig. 4.1 Train speed evolution (Compiled from [29])

The HSR system started its development in the early 1980s, at the end of more than a century and a half of development, during which railway speed had increased more than tenfold, starting at about 50 km/h and reaching about 500 km/h, as shown in Fig. 4.1.

Specifically in Europe, the most significant institutional achievement has been the decision of particular EU (European Union) Member States to build the Trans-European HSR network. The planned length of this network is about 29,000 km, of which 12,500 km will be new lines. The total cost has been estimated to be about €240 Bn (billion), of which €207 Bn has been allocated to HSR static infrastructure and the rest to the HSR rolling stocks. Figure 4.2 shows such development of the HSR network's infrastructure. As can be seen, the length of the network has increased above proportionally until around the year 2003 and is expected to reach 6,000 km in the year 2010 [41].

4.2.2 Recent Development of APT and HSR Traffic

In Europe, both HSR and APT traffic has grown rather intensively during the past decades. Specifically, during the period 1990–2006, APT traffic in EU Member States grew at an average annual rate of 5%, as shown in Fig. 4.3 [1, 41].

During the same period, HSR traffic grew at an average annual rate of 16% (i.e. about three times faster than APT traffic). In particular, the growth of HSR traffic has been strongly influenced by progress in building the HSR infrastructure network, as shown in Fig. 4.2, as well as by other passenger demand-driving forces. Consequently, the most recent forecasts, which do not take into account the impact of the current global economic crisis, indicate that similar growth rates can be expected to take place in the future, and that the volumes of this traffic will double by the year 2015. After the end of the current global crisis, APT traffic volumes are expected to continue to grow at annual rates similar to those in the past (about 5%) [41].

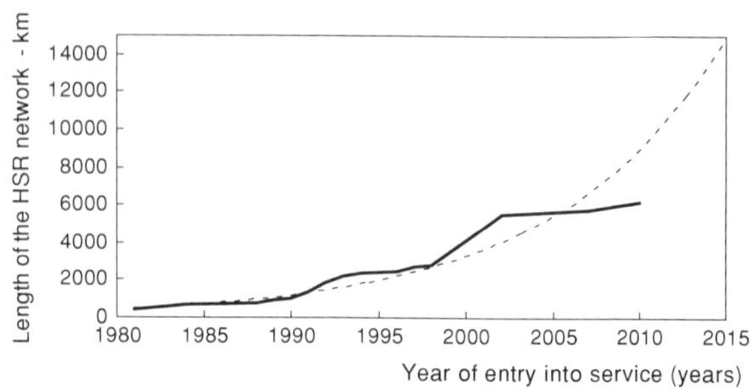

Fig. 4.2 Development of the European HSR network (Compiled from [29, 41])

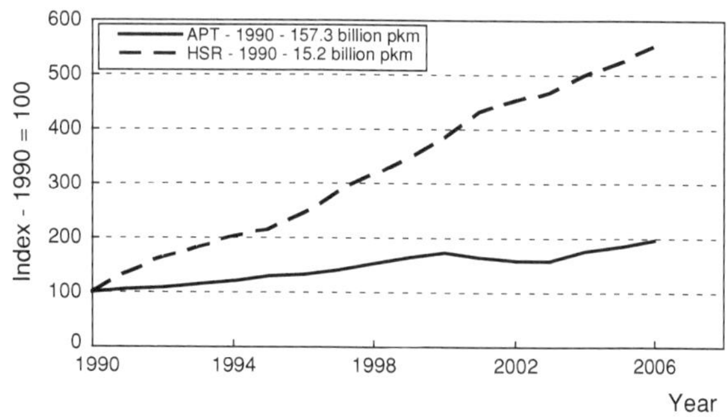

Fig. 4.3 Growth of APT and HSR traffic in the EU(European Union) during the period 1990–2006 (Compiled from [1, 41])

4.2.3 Possible Interactions

In general, the above-mentioned attributes of performance are evaluated partially or fully by the particular actors involved and have caused HSR and APT services operating in the same area to interact with each other in two different ways—competitively and complementarily. Both market relationships between the two modes have been, among other factors, established thanks to connecting particular airports to the HSR network. Table 4.1 shows such development in Europe [5].

4.2.3.1 Competition

In general, competition usually takes place in markets/corridors with substantive volumes of both origin and destination passenger demand, served simultaneously by both modes. In these corridors, passengers choose a particular transport mode on the basis of their perceived generalised out-of pocket travel costs, which are influenced by factors such as the total door-to-door travel time and its cost, fares, and additional costs complementary to the overall quality of service. The alternative with the lower generalised cost is usually preferred. In this case, HSR lines do not need to pass through particular airports at both ends of the competitive markets/corridors [12, 29, 36]. In Europe, competition between HSR and APT takes place in the transport markets/corridors which record substantial volumes of origin–destination (O–D) passenger demand. These include, for example: Madrid–Seville,

Table 4.1 Development of HSR lines to airports in Europe (compiled from [5])

Country	Main routes	First year of service	Airport connection
UK	London–Paris	1994	Full high-speed achieved—2007
	London–Brussels	1994	No direct airport rail connections with London Eurostar Terminal
France	Paris–Lyon	1981	Also serves LYS
	Paris–Bordeaux	1990	All lines connected to CDG rail station
	Paris–London	1994	Conventional (225 km/h) running between Tours and Bordeaux
	Paris–Amsterdam	2009	–
Belgium	Brussels–Paris	1994	No airport connection
	Brussels–London	1994	Full high-speed service to London commenced in 2007
	Brussels–Cologne	2008	Conventional speed running between Liege and the German border
Netherlands	Amsterdam–Paris	2009	Direct connection (at conventional speed) running between Amsterdam Central and Schiphol airport
Germany	Mannheim–Stuttgart	1991	No direct airport connections
	Berlin–Hamburg	1998	No direct airport connections
	Cologne–Frankfurt	2002	CGN and FRA airport connections (Conventional speed running between CGN and Cologne Centre)
	Nuremburg–Munich	2006	No direct airport connections
Italy	Rome–Florence	1991	FCO linked to Rome centre via rail shuttle and metro
	Rome–Naples	2005	MXP connected to Milan centre by Malpensa Express rail service—about 40 min journey
	Turin–Milan	2009	
Spain	Madrid–Seville	1992	MAD connected to city centre (and to the main rail network) by underground service from airport
	Madrid–Malaga	1993	
Sweden	Stockholm–Malmo	1999	Arlanda Express connects ARN with Stockholm city centre in 22 min, for onward "high-speed" connections to Malmo/Gothenburg
	Stockholm–Gothenburg		

Airports: *LYS* Lyon Satolas, *CDG* Charles de Gaulle, *CGN* Bonn-Cologne, *FRA* Frankfurt Main, *FCO* Rome Fiumicino, *MXP* Milan Malpensa, *MAD* Madrid Barajas, *ARN* Stockholm Arlanda

Madrid–Barcelona, London–Paris, London–Brussels, Frankfurt–Cologne, Paris–Marseille, London–Manchester, and London–Edinburgh [36]. Figure 4.4 shows that in the above-mentioned markets/corridors with a travel time of between 1 and 3 h, HSR has taken over a relatively substantial market share from APT, from about 30 to 90% [36].

APT has responded by cancelling most short-haul flights due to their diminishing profitability and/or due to the lack of the convenient slots at some (congested slot constrained) airports at the ends of particular markets/corridors.

Fig. 4.4 Market share of HSR competing with APT depending on the journey time in selected European markets/corridors (Compiled from [13, 36])

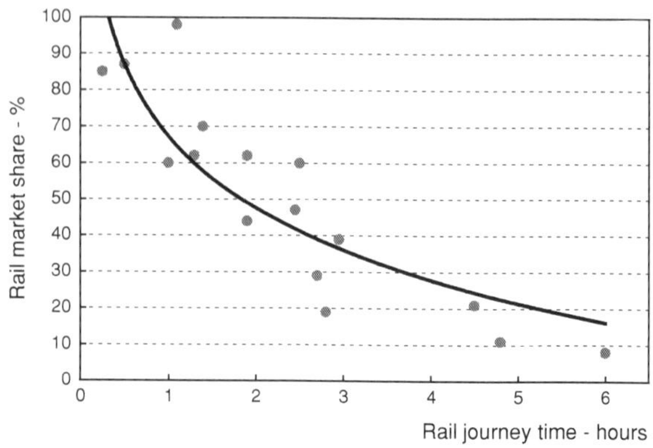

4.2.3.2 Complementarity

In the case of complementarity, air passengers are offered HSR services instead of APT flights at the almost equivalent generalised door-to-door travel costs. In this case, they are ultimately forced to change the mode because it is the only alternative between a given airport and their final destination, and vice versa. In this case, HSR lines have to pass through the given airport [12, 32]. In general, two types of complementary networks may exist, if commercially viable for both HSR and APT operators [26, 30, 41]: (a) HSR may partially or completely replace the APT short-haul "feeder" flights in collecting and distributing passengers between a given hub airport included into the HSR network and particular spoke airports/cities (the passengers' origins and/or destinations). In such cases, the "feeder" HSR services connect to the long-haul flights according to a coordinated timetable. An example in Europe is Frankfurt Main airport (Germany), where some APT short–haul "feeder" flights have been substituted by equivalent HSR services operated in cooperation with the major airline Lufthansa; and (b) the APT system may connect the associated spokes to a given hub airport while HSR may exclusively provide the surface connection between hub airports themselves; as in the case of the HSR connecting Paris CDG and Lyon-Satolas airports in France [16].

4.2.4 Some Social and Environmental Impacts of HS Transport Systems

In considering HSR as a substitute for the APT short-haul flights and consequent contribution to greening, i.e. sustainable development of particular airports, it is useful to analyse some relevant social and particularly environmental performances of both alternatives. The main social performances are congestion and delays, noise, and safety, while environmental performances include energy consumption, related emissions of greenhouse gases, and land use (take).

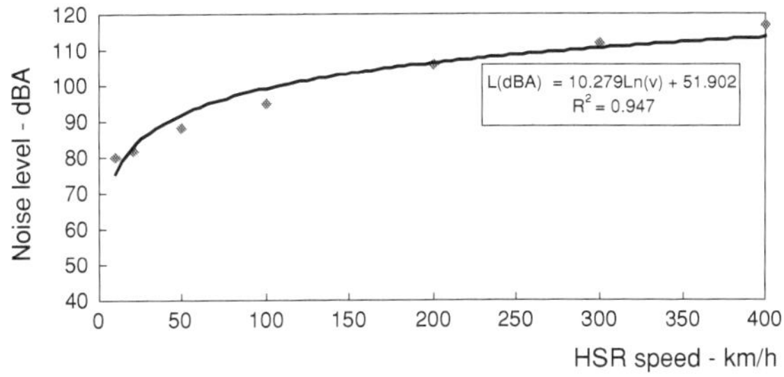

Fig. 4.5 Noise of an HS TGV train depending on cruising speed (Compiled from http://ec.europa.eu/transport/rail/environment/noise.en.htm)

4.2.4.1 Congestion and Delays

Congestion and delays of APT and HSR are elaborated in detail later in this chapter and are therefore not particularly analysed at this point. Nevertheless, it is worth mentioning that APT flights suffer from congestion imposing delays on them, which represent costs for airlines and air passengers. However, HSR services are usually considered to be free of congestion, delays, and related costs for the HSR operators, but their passengers suffer from schedule delays and related costs due to waiting for service.

4.2.4.2 Noise

As mentioned in Chaps. 2 and 3, from the point of view of an affected observer, noise is defined as unwanted sound. In these chapters, APT noise has been elaborated in the form to be used for comparison of the two modes and assessment of the potential effects of their substitution at a given airport.

Noise generated by HSR mostly depends on the technology used. In general, HS trains generate the following noise: wheel-rail noise, pantograph/overhead noise, and aerodynamic noise. Noise is a short time event, impacting the observer while the HS train is passing by. This noise is usually measured in dB(A) (decibels). Measurements of individual noise events have shown that the noise levels differ across different types of HS trains and are positively correlated with the HS trains' cruising speeds, the latter shown for TGV-type trains in Fig. 4.5. This noise includes traction noise, rolling noise, and aerodynamic noise, measured at the right angle distance of 25 m from the track, which is considered as the reference location. As can be seen, this noise generally increases with the train's cruising speed at decreasing rate. In addition, the HSR noise dependent on the distance from the reference location and the train speed can be expressed as follows:

$$L_{HSR}(d,\ v) = 53.938(8.418) - 2.760d(1.585) + 0.192v(3.220)(\mathrm{dB(A)})$$

where $R^2 = 0.936$; $N = 20$; $0 \leq d \leq 9$ km; $0 \leq v \leq 320$ km/h; d is the distance to/from the reference location (km); and v is the speed of the HS train in question(km/h). The above-mentioned relationship does not include the mitigating effects of noise by noise amelioration barriers (walls) if located between the source (HS train) and the observer (i.e. local population). In addition, as intuitively expected, this unaffected noise decreases with distance and increases with cruising speed of the HS trains.

4.2.4.3 Safety

Generally, safety can be defined as the acceptable level of risk of injury, damage and loss of life and/or property. It is measured as the probability of occurrence of an event which may cause the above-mentioned undesirable outcome [27]. Like other transport modes, HSR and APT incidents and accidents burden society and the environment. In particular, incidents usually result in repairable damage to vehicles, property, and non-serious injury. However, accidents also cause loss of life and/or severe injuries, as well as considerable damage and/or complete destruction of vehicles and other property of those directly involved and third parties.

APT system incidents and accidents are rare compared to the number of departures and the volume of p-km (passenger kilometres) during a given period of time. Analysis has shown that global air transport safety has constantly improved over time. For example, during the period 1970–1993, the annual death rate fell from 0.018 to 0.004 (deaths per billion p-km per year) and stabilised around the lower value despite further growth in air traffic volumes. There were 105 deaths resulting from flight operations of European air transport operators over EU territory during the period 1990–1999. Divided by the total cumulative output over the same period of 2012.84 billion p-km, this gives an average annual death rate of $105/(2012.84 \times 9) = 0.0058$ deaths per billion p-km per year [4].

The statistical data on incidents and accidents for HSR indicate that it has also been a very safe transport system. For example, in the EU, HSR has had only one serious accident (German ICE crash at Eschede in 1998, when about 100 people died and 150 were injured). Incidents were more frequent. During the period 1983–2001, the French TGV recorded 12 incidents (train derailment, collision with objects on the track, and terrorist attacks) in which 7 people died and 173 were injured [38]. During the same period, the system carried passengers amounting to 339.5 billion p-km. The annual death rate was $7/(339.5 \times 18) = 0.00114$ deaths per billion p-km per year. These very low figures compared to the volume of output illustrate that both APT and HSR have been very safe systems in terms of annual death rates. Nevertheless, according to the above-mentioned figures, it seems that HSR could be considered slightly safer than APT, but such conclusions always need to be made with caution.

4.2.4.4 Energy Consumption

APT aircraft and HSR trains use different types of energy. This, together with technological diversity, vehicle capacity, and operating conditions (different route lengths and load factors), make their comparison rather complex. However, such comparisons can be made for the short-haul routes where both systems may interact, and under the assumption that one hundred passengers are on-board the vehicles of each system, i.e. an aircraft and an HS train. This provides a unique basis for comparison of the two technologies and partially mitigates the impact of the diversity of operating conditions.

APT aircraft use jet A fuel (kerosene) and/or aviation gasoline. Generally, each burnt kilogram of jet A fuel generates 12.03 kWh of energy [10]. The energy consumption rate (energy–fuel consumption per unit of time) differs for different aircraft types (capacity) and stages of flight as shown in Fig. 2.3. The energy consumption rate is higher for larger aircraft, while for each aircraft type, the consumption rate is higher during take-off and climbing, lower during cruising at optimal altitudes, and the lowest while approaching, landing, and taxiing on the ground.

In order to create a comparable measure for the energy consumption rates of APT as compared to HSR, the following relationship is established:

$ECR_{avg}(APT) = 1000/(vS\lambda)[0.3507S - 2.763]\,(\mathrm{kWh/p\text{-}km})$, where ECR_{avg} (APT) is the energy consumption rate of an aircraft (KWh/p-km); S is the aircraft capacity (seats); v is the aircraft cruising speed (km/h); and λ is the average aircraft load factor ($0 \le \lambda \le 1$). From the above-mentioned expression, for example, the average energy consumption rate of an aircraft with 100 seats and 100 passengers on-board (short-haul flight) amounts to 0.380, the rate for an aircraft with 150 seats (medium-haul) amounts to 0.586, and for an aircraft with 400 seats (long-haul) 1.618 KWh/p-km. The aircraft are assumed to cruise at their fuel-optimal altitudes at a speed of 850 km/h.

HS trains in Europe are powered by electricity. Their marginal energy consumption (quantity of energy per unit of output—KWh/p-km) is mainly proportional to their cruising speed. It is lower during the accelerating/decelerating phase of the trip and higher (but reasonably constant) during cruising at constant speed (of about 250 km/h). Some recent calculations and measurements have shown that French TGV trains (Sud-Est, Atlantique, Reseau and Duplex) consume about 19 KWh/km, which, divided by 100 passengers on-board, gives an average energy consumption rate of 0.19 KWh/p-km (the average train capacity is 430 seats/train). A German ICE train consumes about 22 KWh/km, which divided by 100 passengers gives an average energy consumption rate of 0.22 KWh/p-km [26, 40] (the average train capacity is 380 seats/train).

Based on the above figures and evidence from other studies, the advantage of HSR over ATP in terms of energy consumption rates becomes evident, [29, 30].

4.2.4.5 Emissions of Greenhouse Gases

As mentioned in Chap. 2, emissions of greenhouse gases include gases and particles which are the products of burning fuels needed to power APT aircraft and HS trains.

The marginal quantity of emitted greenhouse gases generated by APT aircraft is much higher than that of HS trains due to the higher energy consumption rates and the types of fuel used (kerosene or gasoline). The dominant gases produced during a flight in terms of quantity are CO_2 and H_2O whose emitted quantities are proportional to the amount of fuel burnt multiplied by the constant factors of 3.18 and 1.21, respectively [2, 24]. As an example, the air pollution rate of CO_2 is calculated as follows: 83 g of kerosene, which is needed to generate 1 kWh of energy, is multiplied by the constant emission rate of kerosene of 3.16 $kgCO_2/kg$ in order to obtain the intensity of emission of CO_2 per unit of energy of 262.6 g/kWh [20]. By multiplying the intensity of emission by the energy consumption rate, the average emission rate of CO_2 can be obtained. For particular aircraft categories, each with 100 passengers on-board, this is as follows: 0.380 KWh/p-km $\times$ 262.6 gCO_2/KWh $=$ 99.8 gCO_2/p-km for an aircraft with100 seats; 0.598 KWh/p-km $\times$ 262.6 gCO_2/KWh $=$ 153.9 gCO_2/p-km for an aircraft with 150 seats; and 1.618 KWh/p-km $\times$ 262.6 gCO_2/KWh $=$ 424.9 gCO_2/p-km for an aircraft with 400 seats. As mentioned in Chaps. 2 and 3, emissions of greenhouse gases by APT are generated locally, around the airports and globally, in the airspace around the aircraft at their cruising altitudes (troposphere) [27, 33].

HS trains use electricity obtained from a combination of different primary sources and the consumption of these sources for electricity production generates greenhouse gas emissions. APT aircraft consume Jet A fuel—kerosene, whose burning generates emissions of greenhouse gases. In both systems, the most important greenhouse gases are CO, NO_x, SO_2, VOCs, CO_2, and PM_{10} (particles). Since CO_2 prevails in the total emitted quantities, it is analysed in more detail, although as mentioned earlier, NO_x and SO_2 have also been considered important greenhouse gases [20, 24].

The quantity of greenhouse gas emissions generated to power an HS train for a given trip depends on the amount of energy consumed and the rate of gas emission from the electricity plants producing the electricity. These plants use different sources for electricity production, and the share of these sources in the total is usually country-specific, [14, 26]. The mixture of sources can include non-renewable (coal, crude oil, natural gas, nuclear) and renewable (wind, solar, hydro, and biomass) sources. Due to the potentially high heterogeneity of the mixture of sources, it is a relatively complex process of estimating the average quantities of emissions of greenhouse gases by HSR. However, some generalisation can be made if, for example, it is assumed that the composition of sources for generating electricity for HSR is the same as that for the country as the whole. Table 4.2 gives an example of such generalisation. As can be seen, due to their slightly higher energy consumption rates and different composition of sources used for generating electricity in Germany and France, respectively, ICE trains have a

Table 4.2 The emission rates of CO_2 depending on the type of source for electricity production and HSR technology

HSR technology/Energy source	Natural gas (1)	Coal (lignite) (2)	Fuel oil (3)	Nuclear/ water/wind (4)	Averages $\sum_{k=1}^{4} p_k APR_k$
TGE (France)					
Proportion of source in gross electricity generation–p_k^a	0.020	0.055	0.019	0.890	–
IOE ($kgCO_2$/KWh)[a]	0.147	0.211	0.345	0	–
ECR (KWh/p–km)[b]	0.190	0.190	0.190	0.190	–
APR = IOE × ECR (gCO_2/p-km)[c]	27.93	40.09	65.66	0	4.011
ICE (Germany)					
Proportion of source in gross electricity generation–p_k^a	0.111	0.501	0.010	0.353	–
IOE ($kgCO_2$/KWh)[b]	0.147	0.211	0.345	0	–
ECR (KWh/p-km)[c]	0.220	0.220	0.220	0.220	–
APR = IOE × ECR (gCO_2/p-km)	32.34	46.24	75.90	0	27.515

[a]Compiled from [14]
[b]*IOE* Intensity of Emissions (converted from [20])
[c]*ECR* Energy Consumption Rate: (quantity per 100 passengers on-board)

much higher average rate of CO_2 emissions than TGV trains. These emissions are considered as indirect emissions spreading from electricity production plants [29, 30].

4.2.4.6 Land Take (Use)

Land take (use) implies taking land for building transport infrastructure. Comparison of APT and HSR with respect to this externality is very complex due to the differing character of both systems' infrastructure. Nevertheless, this can be done on a global scale with respect to the size of land taken to build tracks and runways, rail stations and depots, and entire airports and their ground access systems [3].

For APT, land is primarily taken for building airports. Generally, the size of this land increases with the traffic volumes, which in turn determines the number, length, and configuration of runways. As mentioned in Chap. 3, the typical land increment is about 30 ha/km [23, 31].

The amount of land taken by HSR depends on the length of the line and is not influenced by the volume of traffic. HSR typically records a land increment of 3.2 ha/km for tracks, although some recent research has shown that this increment can be smaller, i.e. about 2.0 ha/km [6–8]). The total land taken by a given HSR line is then roughly proportional to the product of line length and the above constant factor.

In addition, the intensity of land use can sometimes be used for comparing both systems. This can be estimated as the ratio between the volumes of traffic carried during a given period of time and the total land taken. Some estimates have shown that the intensity of land use tends to be comparable for APT and HSR in Europe, i.e. 3.23 and 2.86 million p-km/year/ha, respectively [6, 7]. In addition, in Europe, APT is generally in a favourable position as compared to HSR with respect to new land take. On the one hand, most already built airports only require the incremental increases of land for usually very carefully planned expansion of their airside and landside areas. On the other, new land always needs to be taken to build completely new HSR lines.

4.2.5 Some Potential Effects of APT/HSR Substitution at Airports

The potential effects of substituting some APT short-haul flights with the equivalent HSR services at a given airport can be expressed as savings in the amounts of particular impacts such as airline and air passenger congestion and delays, noise, energy consumption, and related emissions of greenhouse gases, traffic incidents/accidents, land take (use), and their corresponding costs for the particular involved parties. For example, the cost of congestion and delays of particular substituted APT short-haul flights and the costs of those delays, which such flights would otherwise impose on subsequent flights at a given airport during the congestion period, can be considered as savings of the total costs of delays for both airlines and their (air) passengers. With the exception of schedule delay due to waiting for service, HSR en-route delays and related costs can be considered much smaller than those of substituted APT flights. However, air passengers may suffer from schedule delays and related costs due to waiting for the HSR services [32]. In light of the above-mentioned similarity in safety, incidents/accidents and their costs are not particularly considered. This also applies to land take (use), as the very high diversity of the two modes prevents their fair comparison. Nevertheless, some elements may exist if bearing in mind that the HSR lines to/from the airports have always seen some extensions over the already planned lines between particular cities. The eventual savings in the two remaining impacts—noise and energy consumption and related local and global emissions of greenhouse gases—can be particularly relevant for both the local population around a given airport and society as well as the environment, respectively. The noise from APT and HSR has different characteristics at the given airport. For APT flights, it is registered at so-called noise measurement (reference) locations in (and around) the airport area during the aircraft arrival, side-line, and take-off phase of flight. The usual airport-related mitigating measures are described in Chap. 3 [25]. As mentioned above, the noise from HSR increases with the speed of HS trains and spreads along and around HSR lines. However, this noise decreases as the HS trains slow down during their approach to the airport and increases during their acceleration while departing from the airports. Locating HSR lines sufficiently far from populated areas and setting up convenient noise barriers are the usual options for mitigating the noise impact of HSR including that around the airports. In most cases in Europe, HSR lines to and from airports are constructed underground, preventing spreading noise from the HS trains outside the airport area, which would otherwise affect the nearby population [22].

Energy consumption and related emissions of greenhouse gases from APT flights and HSR services also have different characteristics despite the fact that the same greenhouse gases are usually considered. For airports, depending on the scope of consideration, emissions of these gases by APT systems can be expressed in terms of the quantities of CO_{2e} (Carbon Dioxide equivalents) emitted during entire flights (global impact) and/or exclusively during the LTO (Landing and Take-Off) cycles [25]. A similar approach at both the global and local level can be considered for emissions from equivalent HSR services. However, when a certain number of APT flights is substituted by HSR services, both global and local emissions of greenhouse gases need to be considered as saved.

4.2.6 Conditions for Implementing APT/HSR Substitution at Airports

In order to implement APT/HSR substitution at a given airport, some important pre-conditions implying removing or substantive savings in mitigating the existing barriers need to be fulfilled as follows:

1. The given airport must be connected to the HSR network in order to enable APT/HSR substitution;
2. The HSR system needs to provide sufficient capacity to accommodate air passengers transferred from equivalent APT flights. In cases of substitution through competition, HSR operators provide this capacity exclusively, while in cases of complementarity, this capacity is provided through different air/rail code-sharing agreements and alliance partnerships. Some examples in Europe are those between the French national airline Air France, and the national and international rail operators SNCF and Thalys, respectively, as well as between Lufthansa and DB (German National Railways) at Frankfurt Main airport [16, 36];
3. APT/HSR substitution through complementarity is viable only if the generalised travel costs of the air passengers switched from APT to HSR remain very similar, and
4. HSR stations at airports need to be designed and constructed to enable efficient and effective transfer of air passengers and their baggage between the two systems. This primarily applies to the provision of necessary information, realistic walking distances and time, and handling and transfer of baggage.

4.3 Methodology for Assessing Effects of Substituting APT with HSR at an Airport

4.3.1 Background

Research on the relationship between HSR and APT in Europe has mainly focused on proving the operational, economic, and particularly the environmental advantages of HSR compared to APT systems, especially while competing with each other in the short-haul markets/corridors. The findings of such research have been confirmed in practice as HSR has gained a substantial market share in many locations (see Fig. 4.5). However, no exclusive research on the eventual contribution of HSR towards mitigating airport airside congestion and delays, noise, and local and global emission of greenhouse gases by substituting particular APT services/flights through either competition or complementarity has been carried out. In Europe, an exception is the research carried out by EUROCONTROL, which deals with the airport's intermodality in a rather qualitative way [18]. In the US, such research has been conducted, albeit in a quite different context, by (Zhang and Hansen [39]), addressing the optimisation of the total costs of substituting flights, expected to experience long delays at a given hub airport operating under capacity constraints, with coach services. In addition, transferring these affected flights to a reliever hub was considered. Therefore, the main objectives in dealing with this in terms of both research and practical issues in this chapter are as follows:

- Developing a methodology for assessing the potential saving effects in terms of airside congestion and delays, noise, local and global emissions of greenhouse gases, and their costs at a given (congested) airport under given conditions, which could be achieved by substituting particular APT short-haul flights with equivalent HSR services; and
- Carrying out a sensitivity analysis of particular savings with respect to changes of the most influencing factors such as the number of APT flights to be substituted.

The potential savings in other social and environmental impacts, such as traffic accidents/incidents, waste, land take (use), landscape, flora/fauna, and water sources, are not considered.

The methodology is based on the following assumptions:

- The capacity of substitute HSR services at a given airport is sufficient to accommodate air passengers from substituted APT flights;
- The APT aircraft/flight demand including flights to be substituted can be lower, equal, or greater than the airport runway service rate, i.e. capacity;
- APT flights to be substituted by HSR services during a given period of time have similar characteristics such as cost of delay, noise, and local and global air pollution rates; the remaining flights also possess rather homogeneous characteristics, but different from those substituted;

- Congestion and delays, noise, and emissions of greenhouse gases by the ground access systems of the airports and HSR stations at the beginning and end of the corresponding incoming and outgoing APT/HSR routes to/from a given airport, respectively, are not considered (see Chap. 7); and
- The noise and air pollution by serving the potentially substituted APT aircraft/flights at the apron/gate complex of a given airport as well as those at their origin/destination airports are also not considered.

4.3.2 The Structure of the Methodology

The proposed methodology consists of four models: (a) A model for estimating the substitutive capacity of the HSR services; (b) A model for estimating savings of the airline and air passenger delays and related costs; (c) A model for estimating savings of the noise burden and related costs; and (d) A model for assessing savings of emissions of greenhouse gases and related costs.

4.3.2.1 A model for Estimating the Substitutive Capacity of the HSR

The model for estimating the substitutive capacity of HSR services accommodating air passengers transferred from substituted APT services is based on the concept of fully satisfying existing and new volumes of passenger demand. In this case, this may generally require increasing the number of available seats per HSR service (for example by coupling two trains or using a duplex train), increasing the service frequency, and/or both.

Let N_i be the number of APT aircraft/flights to be substituted by equivalent HSR services during the time interval Δt_i. This is the M-th part of the longer period of several hours during the day (τ) (i.e. $i = 1.2, \ldots, M$). Each of these flights carries on average ($S_i\theta_i$) air passengers where S_i is the average aircraft/flight seating capacity and θ_i is the average load factor of a flight. If the proportion of air passengers transferred to HSR services from each APT flight is p_i, the satisfied demand by the HSR implies the following:

$$D_{ik} + \theta_i S_i p_i = F_{ik} C_{ik} \eta_{ik} \tag{4.1a}$$

D_{ik} is the HSR original demand on route (k) in the time interval Δt_i;
F_{ik} is the frequency of the HSR services on route (k) in the time interval Δt_i;
C_{ik} is the seating capacity on an HSR service on route (k) in the time interval Δt_i; and
η_{ik} is the average load factor of an HSR service operating on route (k) in the time interval Δt_i.

Using expression (4.1a), the frequency of HSR services F_{ik} can be determined as follows [28]:

$$F_{ik} = (D_{ik} + \theta_i S_i p_i)/C_{ik}\eta_{ik} \tag{4.1b}$$

where all symbols are as detailed in the previous expressions.

Summing up the frequencies F_{ik} for N_i routes, and then for M time intervals Δt_i, gives the total frequency of HSR services at the airport station, which take part in substituting APT flights over the time period (τ).

4.3.2.2 A Model for Estimating Savings in Congestion and Delays

The model for estimating savings in congestion and delays consists of six sub-models as follows: (1) sub-model of the APT aircraft/flight delays at a given airport; (2) sub-model of the cost of delays of substituted flights; (3) sub-model of the saved cost of delays of flights remaining in the queue; (4) sub-model of the total saved cost of delays by APT/HSR substitution; (5) sub-model of the cost of time of air passengers after switching to HSR services; and (6) sub-model of the total cost savings of airline and air passenger delays.

(i) *The APT aircraft/flight delays at a given airport* The prospective savings of costs of airline and air passenger delays are estimated by the stochastic and deterministic queuing model. The former is applied when APT flight demand is lower or nearly equal to the airport runway service rate, i.e. capacity. The latter is applied when APT demand is greater than the service rate.

(a) *APT aircraft/flight demand is lower than or nearly equal to capacity* In this case, the average delay per APT flight while waiting for arrival at a given airport during the time period Δt_i (i = 1, 2, ..., M) can be estimated with the stochastic queuing model as follows [31]:

$$\overline{w_i(\lambda_i)} = \frac{\lambda_i(\sigma_i^2 + 1/\mu_i^2)}{2(1 - \lambda_i/\mu_i)} \tag{4.2a}$$

where

λ_i is the average arrival rate of APT flights during the time interval Δt_i (aircraft per unit of time);

μ_i is the average APT aircraft/flight service rate during the time interval Δt_i as the reciprocal of the mean service time for arrivals ($\mu_i = 1/\bar{t}_i$; where $\bar{t}_i$ is the minimum average service time per an arrival (time units)), and

σ_i is the standard deviation of the APT flights inter-arrival time (time units).

(b) *APT aircraft/flight demand is greater than capacity* The average delay of an APT aircraft/flight requesting service during the time interval Δt_i can be estimated by the deterministic queuing model applied to the situation shown in Fig. 4.6 [31]: there, the APT aircraft/flight demand exceeds the airport runway capacity for a relatively long period of time (τ) (i.e. for several hours during the day). The cumulative counts of arriving and served demand are represented by curves $A(t)$ and $D(t)$, respectively.

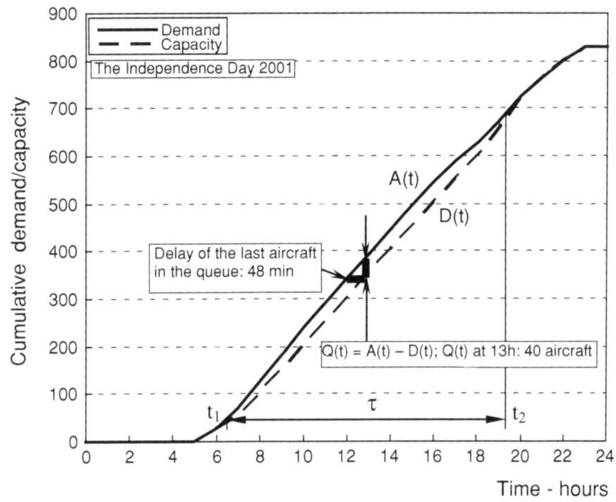

Fig. 4.6 Cumulative count of demand and capacity and resulting congestion and delays at NY LGA Airport (Compiled from [21])

In the particular time interval Δt_i, the derivative of the curve *A*(*t*) with respect to time gives the intensity of demand $\lambda_i = dA[(\Delta t_i)]/dt$. Similarly, the derivative of the curve *D*(*t*) gives the corresponding service rate: $\mu_i = dD[(\Delta t_i)]/dt$, which in the given case is lower than the arrival rate λ_i. From Fig. 4.6, the average delay of an APT aircraft/flight requesting service during the time period (τ) can be estimated as [31]:

$$\overline{w}(\tau) = [1/A(\tau)] \times \int_0^{\tau} [A(t) - D(t)]dt \tag{4.2b}$$

Analogously, the average delay of an APT aircraft/flight during the time interval Δt_i can be estimated as follows [31, 32]:

$$\overline{w_i}(\lambda_i) = \frac{n_{i-1} + 1/2(\lambda_i - \mu_i)\Delta t_i}{\lambda_i} \tag{4.2c}$$

where

n_i is the number of APT aircraft/flights not served in the time interval Δt_{i-}

The other symbols are analogous to those in expression (4.1).

Expressions (4.2a, b, c) can be used in estimating delays of either arriving or departing APT flights to be substituted by incoming and outgoing equivalent HSR services, respectively.

(ii) *The cost of delays of substituted flights* The total delays of N_i APT aircraft/flights to be substituted by the equivalent HSR services during the time interval Δt_i can be estimated based on either expression (4.2a) or (4.2c) as follows:

$$\overline{w_i}(N_i) = \overline{w_i}(\lambda_i)N_i \tag{4.3a}$$

The cost of airline and air passenger delays of the substituted APT aircraft/flights N_i in the time interval Δt_i can be estimated based on expression (4.3a) as follows:

$$C_{1/1i}(N_i) = \overline{w_i}(N_i)\Big[[c_{1i}(S_i) + \alpha_{1i}\theta_i S_i] + [c_{2i}(S_i) + \alpha_{2i}\theta_i S_i]\gamma_i[\overline{w_i}(\lambda_i)]\Big] \tag{4.3b}$$

where

$c_{1i}(S_i)$, $c_{2i}(S_i)$ is the average unit cost of the direct (experienced) and propagated delay, respectively of an APT aircraft/flight of an average seating capacity S_i (€/min);

α_{1i}, α_{2i} is the average value of time of a passenger on-board an APT flight from the group N_i experiencing direct and propagated delays, respectively (€/min-pass); and

$\gamma_i[\overline{w_i}(\lambda_i)]$ is the delay multiplier of an APT flight from the group N_i.

The other symbols are analogous to those in the previous expressions.

The delay multiplier $\gamma_i[w_i\ (\lambda_i)]$ in expression (4.3b) reflects propagation of delays downstream of the daily itinerary of the affected aircraft and air passengers on-board. In general, this multiplier depends on the length of the initial delay. The unit cost of airline delays $c_{1i}(S_i)$ generally depends on the location of realisation of these delays—on the ground before take-off from the origin airport, usually with the engines switched off, or airborne in vicinity of the given (destination) airport. The Difference in the direct and propagated delays of both airlines and air passengers is also suggested, [17]).

(iii) *The saved cost of delays of flights remaining in the queue* The savings of the delays of the remaining aircraft/flights requesting service in the time interval Δt_i is estimated using expression (4.2a) or (4.3b) as follows:

$$\overline{\Delta w_i}(N_i) = [\overline{w_i}(\lambda_i) - \overline{w_i}(\lambda_i - N_i/\Delta t_i)][\lambda_i \Delta t_i - N_i] \tag{4.4a}$$

where all symbols are analogous to those in the previous expressions.

Consequently, the savings in costs of airline and air passenger delays of remaining flights ($\lambda_i\Delta t_i$—N_i) can be estimated as follows:

$$C_{1/2i}(N_i) = \left\{\begin{matrix} \overline{w_i}(\lambda_i)[[c_{3i}(s_i) + \alpha_{3i}\omega_i s_i] + [c_{4i}(s_i) + \alpha_{4i}\omega_i s_i]\gamma_i[\overline{w_i}(\lambda_i)]] \\ -\overline{w_i}(\lambda_i - N/\Delta t_i)\begin{bmatrix} [c_{3i}(s_i) + \alpha_{3i}\omega_i s_i] + [c_{4i}(s_i) \\ + \alpha_{4i}\omega_i s_i]\gamma_i[\overline{w_i}(\lambda - N_i/\Delta t_i)] \end{bmatrix} \end{matrix}\right\}(\lambda_i\Delta t_i - N_i) \tag{4.4b}$$

where

s_i is the average seating capacity of a flight remaining in the queue in time interval Δt_i (seats);

Ω is the average load factor of a flight remaining in the queue in the time interval Δt_i;

$c_{3i}(s_i)$, $c_{4i}(s_i)$ is the average airline cost of direct (experienced) and propagated delay, respectively, of an APT aircraft/flight with an average seating capacity s_i, remaining in the queue during time interval Δt (€/min); and

α_{3i}, α_{4i} is the average value of time of a passenger on-board a remaining flight experiencing direct and propagated delays, respectively, during time interval Δt_i (€/min-pass).

The other symbols are analogous to those in the previous expressions.

As can be seen, the flights remaining in the airport queue can have a different seating capacity and other characteristics such as the cost of delays of both airlines and passengers than the substituted flights.

(iv) *The total saved cost of delays due to APT/HSR substitution* Using expressions (4.3b, 4.4b), the total saved costs of airline and air passenger delays due to substituting N_i APT flights by equivalent HSR services can be estimated as follows:

$$C_{1i}(N_i) = C_{1/1i}(N_i) + C_{1/2i}(N_i) \tag{4.4c}$$

(v) *The cost of time of air passengers after switching to HSR services* The cost of time of air passengers from N_i flights substituted by HSR services during time interval Δt_i, can be estimated as follows [28]:

$$C_{2i}(N_i) = \sum_{k=1}^{N_i} \left(\frac{1}{2F_{ik}} \Delta t_i + \chi_{ik} + \delta_{ik} \right) p_i \alpha_i \theta_i S_i \tag{4.5}$$

where

χ_{ik} is the difference in the average travel time of HSR service and APT flight from the group N_i on route (k) during the time interval Δt_i (time units); and

δ_{ik} is the difference in the average travel time of the ground access systems of APT and HSR used by air passengers from (k)th APT flight from the group N_i during time interval Δt_i (time units).

The other symbols are analogous to those in previous expressions.

In expression (4.5), the first term in parenthesis represents the air passenger schedule delay due to waiting for HSR service. The second term represents the difference in travel time between the two modes. The last term represents the difference in travel time of the ground access systems of both systems (see Chap. 7).

(vi) *The total savings of costs of airline and air passenger delays* The total savings of airline and air passenger costs of delays during the specified period of time (τ) include the savings of costs of direct delays of substituted APT flights and the costs of saved delays of flights remaining in the airport queue, and the costs of air passenger time due to time differences in APT and HSR services. Based on expressions (4.4c, 4.5), the total savings of cost of delays from all APT flights substituted during the time period (τ) can be estimated as follows [32]:

$$C_{1T}(\tau) = \sum_{i=1}^{'M} [C_{1i}(N_i) - C_{2i}(N_i)] \tag{4.6}$$

where all symbols are as in the previous expressions.

4.3.2.3 A Model for Estimating Savings in the Noise Exposure

The model of estimating savings in the noise exposure by substituting particular APT flights with HSR services at a given airport consists of: (a) sub-model for estimating the level of noise exposure to both APT and HSR services; and (b) sub-model for estimating cost of noise exposure, which could eventually be saved by the above-mentioned APT/HSR substitution.

(i) *The level of noise exposure due to APT and HSR* The noise in terms of CSEL (Cumulative Sound Exposure Level) from the prospectively substituted APT flights at a given airport during time interval Δt_i can be estimated as [35]:

$$CSEL_{APT}(\Delta t) = 10 \log_{10} \left[\sum_{k=1}^{N_i} 10^{\mathrm{L_{APT}(k)}/10} \right] \tag{4.7a}$$

where

$L_{APT}(k)$ is the sound exposure level to an aircraft/flight (k) from group N_i (dB(A)).

In expression (4.7a), the variable $L_{APT}(k)$ is estimated either for an arriving or departing flight at the airport where APT/HSR substitution takes place or for the origin and destination airports of the substituted flights. In all cases, it depends on the maximum noise at source and its distance to the "observer". Consequently, if some APT flights are substituted by HSR services, their noise at the corresponding locations will disappear, i.e. will be saved.

In cases when only the local impact at the given airport is considered, HSR noise is irrelevant as mentioned above. However, if considering the wider scale, substituting HSR services creates noise along the tracks outside the airport area, which are predominantly surface constructions. If each HSR service passes several "observers" along its route, the total CSEL along all routes N_i can be estimated as follows [35]:

$$CSEL_{HSR}(\Delta t) = 10 \log_{10} \left[\sum_{k=1}^{N_i} \sum_{l=1}^{F_{ik}} \sum_{m=1}^{f_k} 10^{L_{HAR}(k,l,m)/10} \right] \tag{4.7b}$$

where

$L_{HSR}(k, l, m)$ is the sound exposure level to the (l)-th HSR service passing the (m)-th "observer" located somewhere near the route/line (k) (dB(A)); and

f_k is the number of prospectively annoyed "observers" by HSR noise on route (k).

Consequently, the savings in the total average maximum noise can be estimated as the difference in the values of $CSEL_{APt}(\Delta t)$and $CSEL_{HSR}(\Delta t)$ of both modes calculated by expressions (4.7a, 4.7b). The question, however, remains whether the two noise figures are comparable and to what extent. At the local scale of the given airport where APT/HSR substitution takes place, it certainly makes sense to make such a comparison. On a wider scale, this comparison should be made with caution, if at all. However, in most cases this noise may appear practically irrelevant at both the local level and on the wider scale. Namely, the most exposed "observers" to the noise from both modes are usually properly protected. In particular, those potentially exposed along HSR lines are protected with noise barriers (walls), which are usually considered as a part of the infrastructure and related investment and maintenance costs.

(ii) *Savings in the cost of noise* The above-mentioned generic diversity of the nature of noise created by APT aircraft/flights and their substituted HSR services does not allow us to use expressions (4.7a, 4.7b) to calculate the eventual cost savings under specified conditions. Therefore, the prospective savings in noise exposure by the above-mentioned APT/HSR substitution at a given airport can be estimated more generally as:

$$C_{2T}(\tau) = \max\left[0; \left(\sum_{i=1}^{M} N_i c_{3i/APT} - \sum_{k=1}^{N_i} F_{ik} l_{ik} c_{3i/HSR}\right)\right] \tag{4.8}$$

where

$c_{3i/APT}$ is the average cost of noise of an APT event—aircraft/flight landing or taking-off (€/event); and

$c_{3i/HSR}$ is the average cost of noise of an HS train substituting an APT aircraft/flight (€ct/train-km).

The other symbols are analogous to those in the previous expressions.

4.3.2.4 A Model for Estimating Savings in the Emissions of Greenhouse Gases

The model for estimating prospective savings in the quantity and related costs of local and global emissions of greenhouse gases which could be achieved by substituting some APT flights with equivalent HSR services at a given airport consists of two sub-models: (a) the sub-model for estimating the quantities of emitted greenhouse gases, and (b) the sub-model for calculating savings in the costs of these emissions.

(i) *The quantities of emissions of greenhouse gases* Emissions of greenhouse gases for both APT and HSR can be expressed in terms of CO_{2e} (Carbon Dioxide equivalents), including the above-described air pollutants and their specific characteristics. Thus, the total quantities of greenhouse gases emitted by APT flights substituted by their HSR equivalents at the given airport during the period of time (τ) can be estimated as follows:

$$Q_{APT}(\tau) = e\sum_{i=1}^{M} N_i EC_i(S_i)[(t_i + \overline{w_i}(\lambda_i)] \tag{4.9a}$$

where

- e is the rate of emission of greenhouse gases per unit of consumed energy (tons of CO_{2e}/ton of jet A fuel consumed),
- $EC_i(S_i)$ is the average rate of energy consumption of an APT aircraft/flight with a seating capacity of S_i from the group N_i to be substituted by equivalent HSR services in the time interval Δt_i (tons of jet A fuel/h); and
- t_i is the average duration of an APT flight from the group N_i (h)

The other symbols are analogous to those in the previous expressions.

Similarly, the total emissions of greenhouse gases by HSR services substituting the above-mentioned APT flights can be estimated as follows:

$$Q_{HSR}(\tau) = \sum_{i=1}^{M}\sum_{k=1}^{N_i} F_{ik} EC_{ik} l_{ik} e_{ik} \tag{4.9b}$$

where

- EC_{ik} is the average rate of energy consumed by the HS train substituting an APT flight on route (k) during the time interval Δt_i (kWh/km-train); and
- l_{ik} is the length of HSR service route (k) (km); and
- e_{ik} is the rate of emission of greenhouse gases per unit of consumed electrical energy by HS trains on route (k) during time interval Δt_i (CO_{2e}/kWh).

As mentioned above, HSR services are free of congestion and delays. In addition, the variable e_{ik} in expression (4.9b) is estimated respecting the country-specific emission rates of sources for producing electricity.

(ii) *The savings in the costs of emissions of greenhouse gases* The savings of costs of emissions of CO_{2e} through the above-mentioned substitution of APT with HSR services at a given airport during the time period (τ) can be estimated using expressions (4.9a, 4.9b) as follows:

$$C_{3T}(\tau) = \max\left[0; Q_{APT}(\tau)c_{4/APT} - Q_{HSR}(\tau)c_{4/HSR}\right] \tag{4.10}$$

Where

- $c_{4/APT}$, $c_{4/HSR}$ is the average unit cost of emissions of CO_{2e} by the APT aircraft/flight and of a substituting HSR service, respectively (€/ton of CO_{2e}).

The other symbols are analogous to those in the previous expressions.

Finally, the total savings in airline and air passenger delays, noise, and emissions of greenhouse gases in both quantities and related costs that could be achieved by substituting the given number of APT flights with equivalent HSR services at a given airport during the specified period of time, can be calculated as the sum of expressions (4.6, 4.8, and 4.10).

4.4 Application of Proposed Methodology

4.4.1 Input

4.4.1.1 The Scene

The proposed methodology is applied to calculating the potential savings in the number of and related costs of airline and air passenger delays, noise, and local and global air pollution by substituting some APT short-haul flights with HSR services at London Heathrow airport. The airport in question is still not connected to the European HSR network, although the nearby London City Airport is (Channel Tunnel). Therefore, the results from application of the methodology need to be considered as based on the "what-if" scenario approach. In order to enable flexible APT/HSR substitution through both competition and complementarity, the airport needs to be directly connected to the HSR network implying location of the HSR station in the immediate vicinity as in the case of other large airports—Charles de Gaulle (Paris, France), Amsterdam Schiphol (Amsterdam, The Netherlands), and Frankfurt (Frankfurt Main, Germany) (BAA < 2010). Such development has also been considered to mitigate the current airport airside (runway) congestion, and has seemingly been preferred as an alternative by the current UK coalition Government (which came into power in 2010). The other alternative supported by the former Labour UK Government (in power until the 2010 General Election) implied building a new-third parallel runway for APT short-haul flights. Some other discussions have pointed out that both alternatives need to be in place by the year 2020 to enable further effective and efficient airport growth [11]. Figure 4.7 shows a possible simplified layout of such development. As can be seen, the HSR line is planned to pass through the airport area while connecting London on the east, Cardiff on the west, and Birmingham, Liverpool, and Glasgow on the north-west of the UK. Such layout implies underground construction of the line with a central station located below the passenger terminals, thus enabling efficient and effective transfer of air passengers between the two modes.

Figure 4.8 shows the typical example of the demand-capacity relationship for arrivals at London Heathrow airport before schedule coordination. As can be seen, arrival flight demand exceeds airport capacity during almost the whole day (17 h),

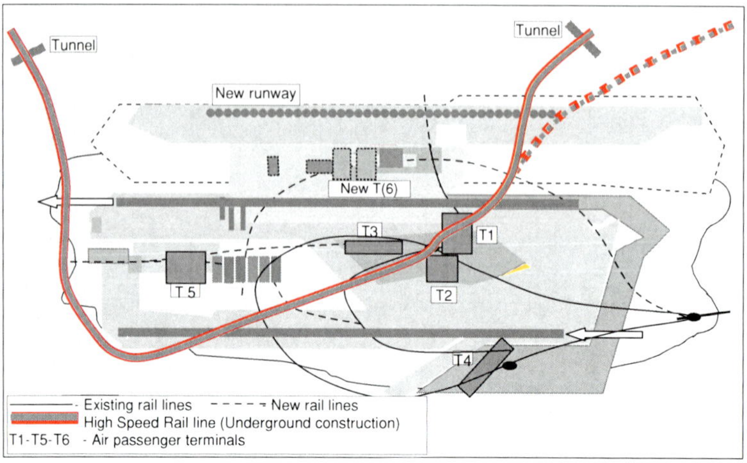

Fig. 4.7 Simplified layout of connecting London Heathrow Airport to the HSR network and its development by 2020 (Compiled from [5])

Fig. 4.8 Typical congestion of arriving aircraft at London Heathrow Airport (UK) before schedule coordination (a peak summer day in 2008) (Compiled from [32])

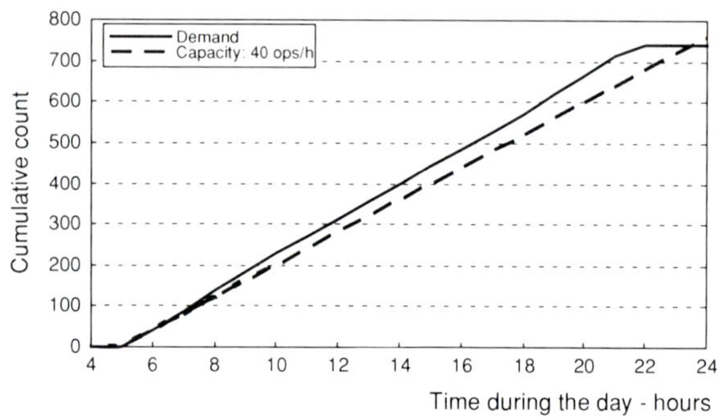

indicating conditions of severe congestion and delays. Flight demand is currently accommodated at one of the two parallel runways operating in the "segregated" mode (i.e. one exclusively serving arrivals and the other exclusively departures). The service rate, i.e. capacity, for arrivals (and departures) under good weather conditions is: $\mu_{i(.)} = 40$ ops/h.

The airport's service rate or capacity consists of 2.6% small, 56.7% medium, and 40.7% large aircraft.

In general, all short-haul APT flights carried out by small and some of those carried out by medium-sized aircraft could be reasonable candidates for substitution with HSR services. These are mainly UK domestic flights and those between the UK and the north of France and Benelux countries (the Netherlands, Belgium, and Luxembourg). The daily number of these arriving and departing fights, each lasting about 1.6 h, is around 110/110. This is about 15% of the total number of daily scheduled flights at the airport. Regarding the comparative door-to-door time of 1–3 h, which could be achieved by comparable HSR services along the particular routes, it is reasonable to expect that up to 50-60% of the above-mentioned short-haul flights could be substituted with HSR services through competition (see Fig. 4.4). The rest could eventually be substituted through complementarity, but very likely only after the HSR network spreads from the airport to the other parts of the UK [5, 34].

4.4.1.2 Characteristics of APT Flights and HSR Services

Consequently, in the application of the methodology, the average length of all routes of both APT flights and equivalent HSR services is adopted to be: $l_{i/APT} \approx l_{i/HSR} = 750$ km, which with an average block speed of $v_{i/APT} = 475$ km/h gives flight duration of $t_{i/APT} = 1.6$ h. The APT short-haul flights are carried out by aircraft of an average seating capacity of $S_{i/APT} = 130$ seats and load factor of $\theta_{i/APT} = 0.673$. The remaining flights in the queue are of the average seating capacity of $s_{i/APT} = 220$ seats and load factor: $\omega_{i/APT} = 0.60$.

The equivalent HSR services are carried out along the routes of length of $l_{i/HSR} = 750$ km at an average speed of $v_{i/HSR} = 270$ km/h, which gives an average travel time of $t_{i/HSR} = 2.8$ h. The seating capacity of an HS train is adopted to be: $C_{i/HSR} = 485$ seats and load factor $\eta_{i/HSR} = 0.60$ [13, 36]. The number of passengers already on-board each HSR service taking over air passengers from the substituted APT flights at the airport is adopted to be: $D_{ik/HSR} = 300$ m which seems reasonable if the time in which the APT/HSR substitution takes place is adopted to be $\tau = \Delta t_i = 1$ h.

The difference in the travel time by the two modes is equal to: $\chi_{i/APT/HSR} = 1.2$ h. This is however compensated by the difference in the ground access time δ, which is usually positive for HSR. Since APT/HSR substitution is assumed to take place exclusively through complementarity, the proportion of air passengers switching to the HSR is adopted to be $p_i = 1.0$ ($I = 1, 2, \ldots, M$).

4.4.1.3 Congestion and Delays

The average cost of airline direct delay of the substituted APT flights is adopted to be: $c_{1i}(S_i) = 47$€/min and of their propagated delay: $c_{2i}(S_i) = 62$€/min. The corresponding cost of flights remaining in the queue and benefiting from such substitution are adopted to be: $c_{3i}(s_i) = 78$€/min and $c_{4i}(s_i) = 101$€/min, respectively. The average cost of air passenger direct and propagated delays of substituted APT flights is adopted to be: $\alpha_{1i//APT} = \alpha_{2i//APT} = \alpha_{3i//APT} = \alpha_{4i//APT} =$ 69€/h [17].

The delay denominator is synthesised for three different periods of the day when substitution is likely to take place, i.e. in the early morning, early afternoon, and in the late afternoon, as follows [32]:

$\gamma_1[\overline{w_1}(\lambda_1)] = 1.391e^{0.0073\overline{w_1}(\lambda_1)}$; $R_1^2 = 0.967$; (Morning—08:00);
$\gamma_2[\overline{w_2}(\lambda_2)] = 1.1851e^{0.0058\overline{w_2}(\lambda_2)}$; $R_2^2 = 0.982$; (Early afternoon—13:00); and
$\gamma_3[\overline{w_3}(\lambda_3)] = 1.0509e^{0.0037\overline{w_3}(\lambda_3)}$; $R_3^2 = 0.994$; (Late afternoon—18:00).

These delay denominators indicate that a stronger network impact can be expected from longer morning initial delays due to the knock-on effect to a greater number of flights of the affected aircraft's daily itineraries. The impact is lower later in the day with reducing prospectively affected flights in the aircraft's daily itinerary. Since the HSR services are assumed to be free of congestion and delays, the corresponding costs are assumed to be zero.

4.4.1.4 Noise

The noise effects are considered at the local (airport) scale. Thus, the noise of an APT aircraft/flight in terms of SEL (Sound Exposure Level) measured at the airport noise measurement location for arrivals (2 km from the landing threshold on the extended runway centreline) is adopted as relevant. For aircraft types B737-300/400 and A319/313/320, this noise level amounts to: $L_{APT}(k, l, m) = 92$ dB(A) and is attenuated with increasing distance to an "observer" [32]. The corresponding cost of noise is adopted to be: $c_{3i/APT} = 61$€/event (i.e. an arrival or a departure) [19].

The noise of HSR services operating within the airport area is estimated from the above-mentioned causal relationship. The distance from the track is adopted to be $d_{HSR} = 25$ m (width of the tunnel) and the average speed $v_{HSR} = 90$ km/h (within the airport area). The unit cost of noise is adopted to be: $c_{3/HSR} = 2.57$€ct/Train-km [9].

4.4.1.5 Emissions of Greenhouse gases

The fuel consumption of APT short-haul flights to be substituted by HSR services is adopted to be: $EC_{i/APT} = 3.33$ tons/flight. The rates of emissions of greenhouse gases are: $e_{1/APT} = 3.18$ kgCO_2, $e_{2/APT} = 1.21$ kgH_2O, and $e_{3/APT} = 0.84$ gSO_2/kg of jet A fuel, which gives a total emission rate of CO_{2e} of: $e = 4.39$ kgCO_{2e}/kg of Jet A fuel. The average cost of air pollution including both local and global impact is: $c_{4/APT} = 43.6$€/ton of CO_{2e} [19].

The HSR services are assumed to be carried out by TGV-type trains consuming electricity at an average rate of: $EC_{ik} = 21.825$ kWh/train-km [33]. The average rate of emissions of CO_{2e} from electricity production in the UK, France, Belgium, and the Netherlands is estimated to be: $e_{ik} = 0.452$ kgCO_{2e}/kWh [15]. The average cost of emissions of greenhouse gases by an HSR service including both local and global impact is adopted to be: $c_{4/HSR} = 26.7$€ct/train-km [9].

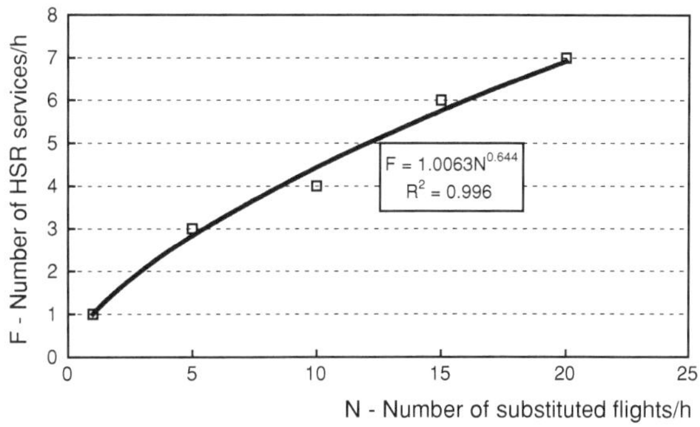

Fig. 4.9 The HSR substituting capacity depending on the number of APT substituted flights

4.4.2 Analysis of Results

4.4.2.1 The Substitutive Capacity of HSR

Figure 4.9 shows the relationship between the number of substituted APT flights and the equivalent HSR services in a given period of time (1 h) [32].

As can be seen, the number of these HSR services increases with the number of APT flights to be substituted at a decreasing rate. In addition, this indicates that the HSR substitutive capacity is not likely to act as a barrier to APT/HSR substitution under given conditions.

4.4.2.2 Congestion and Delays

Figure 4.10a, b, c, d shows savings of the particular categories on flight delays and related costs depending on the number of substituted APT flights and the time of day when substitution takes place.

Figure 4.10a shows that the savings of direct airline delays increase with the number of substituted APT flights at an almost linear rate. These delays increase with the initial queue and related delays with which the flights to be substituted are confronted with.

Consequently, in the given example, the longest airline delays are saved by flights substituted in the late afternoon (faced with an initial queue of 52 flights/aircraft) while the shortest delays are saved by flights substituted in the morning (faced with an initial queue of 18 aircraft/flights).

Figure 4.10b shows that the savings of delays of flights remaining in the queue also increase with the number of substituted flights. In this case, these savings are the highest when substitution takes place in the morning and the lowest when substitution takes place in the late afternoon, for any number of substituted flights. This occurs due to the fact that in the given example, the number of potentially

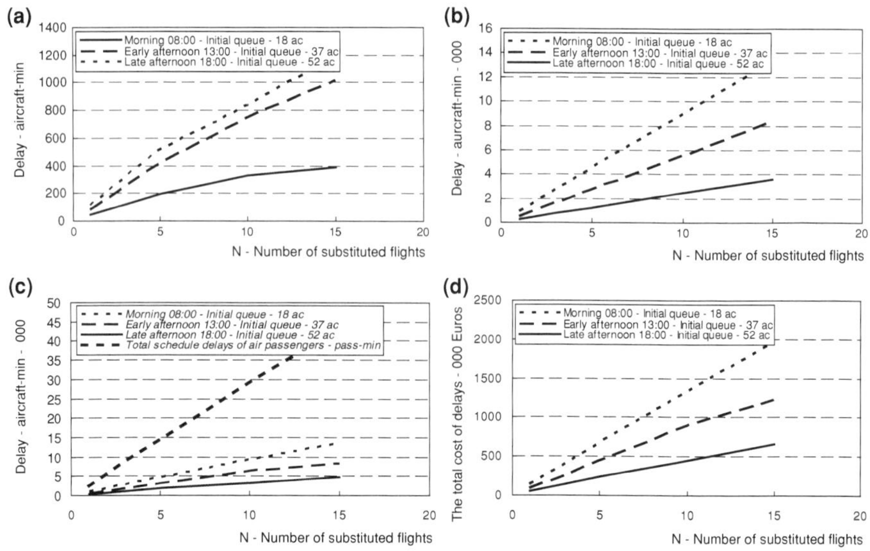

Fig. 4.10 Savings in airline and air passenger delays and costs depending on the number of substituted APT flights and time of day in the given example (Compiled from [32]) . **a** Direct airline delays, **b** delays imposed on the remaining flights in the queue, **c** total airline delays and air passenger schedule delays, **d** total airline and air passenger cost of delays

delay-relieved flights scheduled to arrive after the substituted flights, is highest in the morning and lowest in the late afternoon.

Figure 4.10c shows that the savings of the total delays of both substituted and remaining flights also increase with the number of substituted flights, regardless of the time of day. Since the remaining flights dominate, the highest savings of total delays could be achieved by substituting flights in the morning and the lowest by substitution in the late afternoon. At the same time the total schedule delays of all air passengers due to waiting for HSR services and due to the differences in the service time by both modes increase with the number of substituted flights independent of the time of day. In the given example, in the absolute sense, they appear to be much higher than that of the substituted flights.

Figure 4.10d shows the total cost of airline and air passenger delays as savings from APT/HSR substitution in the given example. As can be seen, these savings are substantial and increase with the number of substituted flights. They are also the highest for substitution in the morning and the lowest for substitution in the late afternoon. The additional calculations show that congestion and delays in Fig. 4.8 would disappear if the APT short-haul flights uniformly scheduled during the day, were substituted at a rate of 5 flights/h. This implies about 85 substituted flights over the congestion period of 17 h, which is around 70% of the present number of daily short-haul flights.

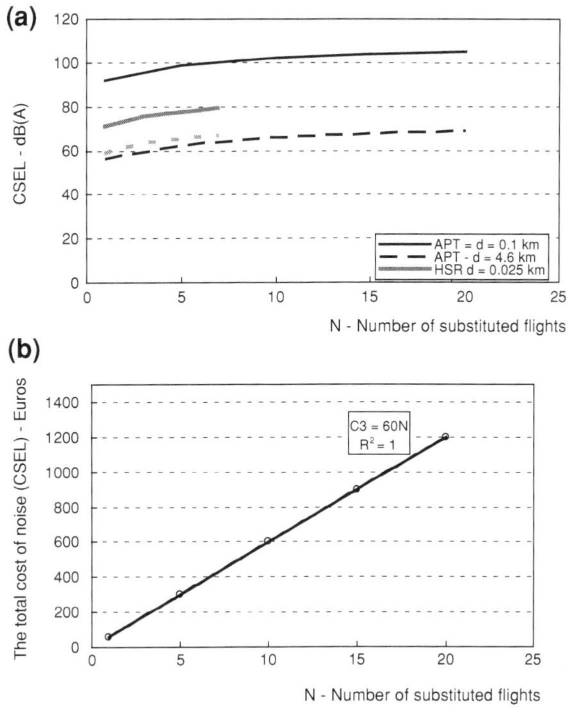

Fig. 4.11 Savings in sound exposure and related costs depending on the number of substituted APT flights in the given example. (Compiled from [32]) **a** Cumulative Sound Exposure Level (CSEL), **b** cost of saved CSEL

4.4.2.3 Noise

Figures 4.11a, b show the savings in the cumulative sound exposure level and related costs in the given example.

Figure 4.11a shows the cumulative sound exposure level (CSEL) for both APT and HSR. As can be seen, CSEL by APT increases with the number of flights at a decreasing rate and decreases with their distance from an "observer". The same happens with CSEL by HSR, which, absorbed within the tunnel through the airport area, is given only for comparative purposes.

This noise is lower than that of the APT aircraft/flights due to the lower noise of the slower moving HS trains and their lower service frequency.

Figure 4.11b shows that, under the above-mentioned conditions, the savings in the cost of noise burden originate and increase exclusively in line with the number of substituted APT flights. In absolute terms, these savings are much smaller as compared to those of the airline and air passenger delays.

4.4.2.4 Emissions of Greenhouse Gases

Figures 4.12a, b show the quantities of emissions of CO_{2e} and related costs as savings due to APT/HSR substitution in the given example.

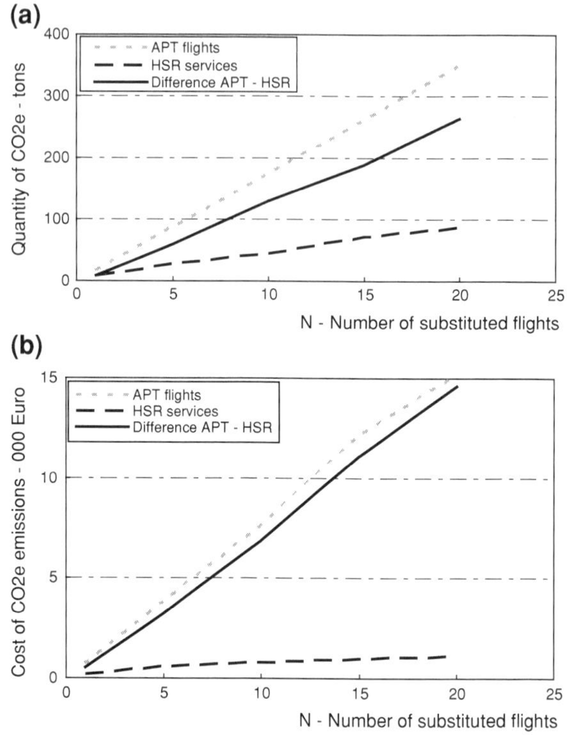

Fig. 4.12 Savings in CO_{2e} emissions and related costs depending on the number of substituted APT flights in the given example. (Compiled from [32]) **a** Quantity of CO_{2e} emissions, **b** cost of CO_{2e} emissions

Figure 4.12a shows that the emissions of greenhouse gases of APT flights are much higher than those of HSR services. Thus, savings are positive and linearly increase with the number of substituted APT flights ranging from ten to three hundred tons of CO_{2e}.

Figure 4.12b shows that the costs of the above-mentioned emissions of both modes and their differences considered as savings also increase with the number of substituted APT flights. The savings range between one and fourteen thousand Euros, and thus are much greater than those from noise but again substantially lower than those from the airline and air passenger delays.

Summing up, the savings in the particular impacts give the total savings in the given example. Nevertheless, it needs to be mentioned that these savings relate to substitution of a very limited number of the APT flights scheduled during one hour of the given day. Thus, for example, if all APT short-haul flights in Fig. 4.12 were substituted, the total daily savings of the above-mentioned impacts would be higher.

4.5 Concluding Remarks

This chapter has described the potential social and environmental effects, which could be achieved by developing a large congested airport into a true multimodal transport node, and consequently substituting particular APT (Air Passenger

Transport) short-haul flights by equivalent HSR (High-Speed Rail) services. The effects have been expressed in terms of savings of the airline and air passenger congestion and delays, noise, and emissions of greenhouse gases, and their related costs. An extensive analysis of the social and environmental performances of both systems has been used as the basis for developing a methodology for assessing the savings in the above-mentioned impacts. The methodology has been applied to a large congested European airport (London Heathrow, UK), which still needs to be connected to the UK and European HSR network. The results from this application have indicated that, in the given example, HSR can provide sufficient capacity for handling air passengers from substituted APT flights. Very modest substitution of APT flights (up to 2%) with equivalent HSR services could have substantial daily saving potential (up to about 20% of delays and up to 17% of their costs). In general, these savings would increase if substitution of a larger number of APT short-haul flights took place earlier in the day. Savings in the noise burden and the emissions of greenhouse gases in terms of CO_{2e} and their costs could also be achieved under the given circumstances. In absolute terms, the savings in the cost of noise are much lower than those of the emissions of greenhouse gases, but both appear to be much lower than those of the airline and air passenger delays. In any case, the above-mentioned savings generally increase with the number of substituted APT flights. Consequently, they can certainly contribute to greening, i.e. sustainable development of the airside area of the given airport. In addition, both the methodology and results could be used by local and national policy makers, airlines, airport and HSR planners and designers as inputs for evaluating the overall socio-economic feasibility of developing an airport into a true multimodal transport node by including it into the HSR network. This could also be an alternative for increasing the airport runway capacity without building new runways using additional land.

In any case, care should be exercised due to the controversy and complexity in estimating particular impacts such as (specifically) noise by both systems/modes at the given airport. Therefore, the results from the given example based on the "what-if" scenario approach can be considered only as an orientation for further more detailed and refined research. Nevertheless, it is quite clear that some social and environmental benefits and consequent contribution to greening, i.e. sustainable development of the given airports into true multimodal transport nodes under the described circumstances can be achieved.

References

1. AEA (2008) Yearbook 2000–2006. Association of European airlines, Paris, France, 2007
2. Archer LJ (1993) Aircraft emissions and environment. Oxford Institute for energy studies, Oxford, UK EV 17
3. ATAG (1996) The economic benefits of air transport. Air transport action group, Geneva, Switzerland

4. ATAG (2000) European air traffic forecast 1980–2010. Air transport action group, Geneva, Switzerland
5. BAA (2010) High-speed rail access to Heathrow: BAA submission to lord mawhinney report. British airport Authority, London, UK
6. CEC (1993a) European high-speed rail network: Socio-economic impact study. Commission of the European Communities, and Directorate general for transport, Final Report, Brussels, Belgium
7. CEC (1993b) The European high-speed train network: environmental impact assessment, executive summary. Commission of the European Communities, Directorate general for transport, Brussels, Belgium
8. CEC (1995) High-speed Europe. High level group 'the European high speed train network'. Commission of the European Communities, Directorate general for transport, Brussels, Belgium
9. CE Delft (2008) Handbook on estimation of external costs in the transport sector, (produced within the study internalization measures and policies for all external cost of transport (IMPACT)). CE Delft-Solutions for Environment Economy and Technology Delft, The Netherlands
10. Chevron (2002) Aviation fuels, Chevron Product Corporation. http://www.chevron.com/prodserv/fuels
11. DfT (2009) Britain's transport infrastructure—adding capacity at Heathrow: decisions following consultation. DfT publications, department for transport, London, UK
12. EC (1998) Interaction between high-speed rail and air passenger transport. Final Report, Action COST 318, European Commission, DG VII, Luxembourg
13. EC (2002) Developing EU (International) rail passenger transport. OGM-Final Report European Commission. Directorate General Energy and Transport, Brussels, Belgium
14. EC (2004) Statistical yearbook: transport–2004. European Commission, Luxembourg
15. EC (2009) EU energy and transport in figures, statistical pocket book. Directorate General for Energy and Transport, Brussels, Belgium
16. ECMT (2003) Airports as multimodal interchange nodes. Conclusion of Round Table 126, European Conference of Ministers of Transport, Paris Cedex 16, France, p 4
17. EEC (2004) Evaluation of true cost to airlines of one minute of airborne or ground delays. Westminster University (UK). Performance Review Unit, EUROCONTROL, Brussels, Belgium
18. EEC (2005) Potential airport intermodality development. WP 2.1.1 DOC, EUROCONTROL, Brussels, Belgium
19. EEC (2007) Standard inputs for EUROCONTROL cost benefit analysis. EUROCONTROL, Belgium Brussels
20. EIA (2002) Voluntary reporting on greenhouse gases: Form–1605. Energy information administration. U.S. Department of Energy, Washington, DC
21. FAA (2007) FAA operations & performance data, Department of Transportation. Federal Aviation Administration. http://www.apo.data.faa.gov/, Washington DC, USA
22. HMMH (1996) Summary of european high-speed rail noise and vibration measurements. HMMH Report No. 293630-2, Miller & Harris Inc., Burlington, UK
23. Horonjeff R, McKelvey FX (1994) Planning & design of airports, 4th edn. McGraw Hill Inc., New York
24. ICAO (1993a) Aircraft engine emissions, environmental protection. Annex 16, vol 2. International Civil Aviation Organisation, Montreal, Canada
25. ICAO (1993b) Aircraft noise, environmental protection. Annex 16, vol 1. International Civil aviation Organization, Montreal, Canada
26. ITA (1991) Rail/Air complementarity in Europe: the impact of high-speed rail services. Institute du Transport Aèrien, Paris, France
27. Janic M (1999) Aviation and externalities: the accomplishments and problems. Transp Res D 4:159–180

28. Janic M (2000) Analysis and modeling of air transport system: capacity quality of services and economics. Gordon and Breach Science Publisher, Amsterdam
29. Janic M (2003) Multiple criteria evaluation of High-Speed Rail, TRANSRAPID MAGLEV and air passenger transport systems in Europe. Transp Plann Technol 26(6):491–512
30. Janic M (2003b) High-Speed Rail and air passenger transport: a comparison of the operational environmental performance. Proc Inst Mech Eng F J Rail Rapid Transit 217(4): 259–269
31. Janic M (2009) The airport analysis, planning, and design: demand, capacity, and congestion. Nova Science Publishers, Inc, New York, USA
32. Janic M (2010) True multimodalism for mitigating the airport congestion: substitution of air passenger transport by high-speed rail. Transportation Research Record (TRR) (forthcoming)
33. Lukaszewicz P, Anderson E (2009) Green train energy consumption: estimation of high-speed rail operations. GRONATAGET, KTH Railway Group, Stockholm, Sweden
34. RMT (2009) Who says there is no alternative? An assessment of the potential of rail to cut air travel. Rail and maritime transport (Union of Workers), London, UK
35. Ruijgrok CJJ (2000) Elements of Aviation Acoustics. Het Spinhuis Publishers, Amsterdam, The Netherlands
36. SDG (2006) Air and rail competition and complementarity. Steer Davies Gleave. Study prepared for European Commission DG Energy and transport, London, UK
37. Sean B (2005) The Role of Airports in the Transport Chains. ECMT (European Conference of Ministers of Transport), Round Table 120, Paris, France
38. TGVweb (2002) High-Speed Rail Safety: TGV Accidents. http://www.mercurio.iet.unipi.it/tgv
39. UIC (2002) Shaping the Railway of the 21st Century: AIR/HSR from Competition to Complementarity. Presentation. http://www.uic.asso.fr/d, International Union of Railways, Paris, France
40. UIC (2008) Fast Track & Sustainable Mobility. 5th World Congress on High Speed Rail, International Union of Railways, 17–19 March, Amsterdam, The Netherlands, p 10
41. Zhang Y, Hansen M (2007) Real Time Inter-Modal Substitution (RTIMS) as an Airport Congestion Management Strategy. 87th Transportation Research Board (TRB) Conference, Washington DC, USA

Chapter 5
Greening the Airport Airside Area I: Increasing Runway Capacity Without Increasing Airport Size

5.1 Introduction

The parties involved in the air transport system such as airport operators, airlines, and ATC (Air Traffic Control), the systems' regulatory bodies at both national and international level, planners and researchers (academic and consultants) have made great efforts to provide sufficient airport runway capacity to adequately serve the continually growing demand. However, these efforts have had very limited success or have even, in many cases, been unsuccessful. In addition to the growing air transport demand, the specific environmental (mainly noise) and particularly the land take (use) constraints at many large airports both in Europe and the US have prevented the full utilisation of the designed airport runway "ultimate" and "practical" capacities. The former is defined as the maximum number of atms (air transport movements) carried out at a given runway system under conditions of constant demand for service during the specified period of time, while the latter is defined as the maximum number of atms per period of time, which enables maintaining the average delay per atm within the prescribed limit(s) (one atm is one landing or one take-off). Under the continuously growing demand, these capacity constraints have caused an increasing imbalance between demand and capacity, which has increased airport airside congestion, delays, and related costs for airlines, air passengers, and air cargo shipments.

Under such circumstances, the question arises whether it is possible to increase the capacity of existing runways by deploying advanced operational regulations and procedures supported by existing and advanced technologies, and thus achieve capacity gains without requiring additional land for building new runway(s). This will actually enable greening, i.e. more sustainable development of given airport(s) at least in the medium-term with respect to any additional land requirements (see Chap. 3) [4].

This chapter describes the potential of some of the above-mentioned advanced operational regulations and procedures for increasing the landing capacity of a single runway such as: (1) ATC time-based separation rules; (2) prioritising

M. Janić, *Greening Airports*, Green Energy and Technology,
DOI: 10.1007/978-0-85729-658-0_5, © Springer-Verlag London Limited 2011

landings; and (3) ATC vertical distance-based separation rules. The term (acronym) ATC is used in this chapter instead of ATC/ATM since we are dealing with ATMs (air traffic movements) at the operational and tactical level.

5.2 Advanced Technologies for Increasing Runway Landing Capacity

The development and implementation of particular advanced operational regulations and procedures such as ATC time-based instead of the distance-based separation, prioritising instead of currently used FCFS (First-Come–First-Served) priority rules for landing aircraft, and steeper approach procedures for landings on closely spaced parallel runways requires improvements of existing and development of new advanced technologies. Both approaches are currently under development as part of the US NextGen (Next Generation Air Transport System) and European-EUROCONTROL SESAR (Single European Sky ATM Research) research and development programs, but most research relating to the latter is still at the conceptual or initial test level [9, 11, 16–18]. Some advanced technologies which could be applied to the aircraft landings in any given runway configuration are given in Table 5.1.

As indicated, implementation of most of the above-mentioned technologies is expected to take place in the medium to long term, i.e. by the year 2015–2020, and beyond. In particular, these technologies are expected to significantly improve the quality and reliability of collection and updating of data on the aircrafts' position in airspace, vastly improved capability to follow planned 4D trajectories, and consequently reducing or completely eliminating existing ATC time and distance-based separation buffers. These technologies will also enable reducing separation and increasing diversity of ATC separation rules to be applied safely to landings on a single runway. Thus, their contribution to increasing the runway's landing capacity is expected to be relatively substantive. In addition, these technologies will enable sharing the responsibility for separation between the ATC controllers and pilots (currently, except in situations of conflict resolution, full responsibility for establishing and maintaining separation between aircraft lies with ATC controllers). In parallel, the various ATC ground decision-support tools, some of which are aimed at contributing to more efficient and effective matching of demand to capacity at both tactical and operational levels, while maintaining the ATC controllers' workload under the prescribed limits, are being deployed. Specifically, in Europe and the US, some of these existing tools include GHP (Ground Holding Program), ATFM (Air Traffic Flow Management) in Europe and AFP (Airspace Flow Program) in the US, as well as FSM (Flight Schedule Monitor), FSA (Flight Schedule Analyzer) and TMA (Traffic Management Advisor). Nevertheless, innovative ATC tactical and operational decision support tools, in addition to CTAS and Arrival/Departure Manager, will need to be developed or existing systems modified and upgraded [9].

Table 5.1 Advanced technologies supporting advanced landing procedures

Location	Technology	Availability
Air traffic flows management tools—used by ATC	CTAS (Centre/TRACON Automation System) assists in optimising the arrival flow and runway assignment	Immediate
	Integrated Arrival/Departure Manager enables planning in advance and updating the arrival sequences including replacement of the FCFS (First-Come–First-Served) rule with another sequencing rule implying prioritising particular aircraft wake-vortex categories and use of the time-based instead of the distance-based separation rules also between landing aircraft	Immediate, but improvements required
Air traffic surveillance equipment—used by ATC on the ground	Improved precision RADAR enables reduction of the minimal separation between aircraft from 3 to 2.5 nm	Immediate with additional improvements in the medium term
	PRM—Precision Runway Monitor consisting of a beacon radar and computer predictive displays enables the independent use of dual- and triple-dependent parallel runways spaced less than 4,300 ft apart	Immediate
	Terminal Wake Vortex Detection System provides information about the wake-vortex behaviour during landing and take-off	Medium to long term
Improved aircraft on-board avionics	FMS 4D—Flight Management System enables more precise maintenance of the time schedule according to the flight plan, which reduces the position error of arrivals at the final approach gate	Immediate
	Sending 4D trajectory enables exchanging the continuously updated aircraft current and expected position obtained from FMS to the external recipients such as other aircraft-pilots and ATC controller(s) on the ground	Long term
	ADS-B—Automatic Dependent Surveillance Broadcasting improves situation awareness both on-board and on the ground and is used independently, but in addition to TCAS and enhanced CDTI (Cockpit Display of Traffic Information)	Medium term
	CDTI—Cockpit Display of Traffic Information provides integrated traffic data on-board the aircraft, which may reduce separation rules between aircraft	Medium to long term

(continued)

Table 5.1 (continued)

Location	Technology	Availability
	TCAS—Traffic Alert and Collision Avoidance System shows the spatial relation of two aircraft and provides instructions to avoid potential conflicts	Immediate
	ACAS—Airborne Collision Avoidance System strengthens the quality of information for ASAS thus enabling replacing PRM (above) and reducing position errors in all three dimensions	Medium to long term
	ASAS—Airborne Separation Assistance (Assurance) System enables airborne surveillance, display of traffic information, and consequently sequencing and merging based on the data from ADS-B	Long term
	WVDS (Wake Vortex Detector System) on-board the aircraft enables collecting and displaying information about the existing wake vortex to both pilots and ATC controllers	Medium to long term
	WAAS—Wide Area Augmentation System improves basic GPS accuracy both horizontally and vertically	Medium term
	AILS—Airborne Information for Lateral Spacing improves the navigation precision while approaching closely spaced parallel runways	Medium term
	LVLASO—Low Visibility Landing and Surface Operating Program reduces, controls and predicts the runway occupancy time	Medium term
"Mixed" traffic surveillance and conflict alert equipment— used by ATC and on-board the aircraft	Distributed Air Ground Solution combines ADS/B, TCAS, and Free Flight devices enabling simultaneous aircraft-ATC traffic surveillance, alerting and resolution of potential conflicts	Medium term

Compiled from: [8, 9, 11, 21]

5.3 Advanced Procedures and Regulations for Increasing Runway Landing Capacity

5.3.1 The ATC Time-Based Separation Rules

5.3.1.1 Background

The ATC time-based instead of the distance-based separation rules between landing aircraft have been studied as an option for stabilising variations of the single runway landing capacity caused by variations of the aircrafts' immediate and final approach speeds due to changing head winds. Under such circumstances, the time separation rules based on the ATC minimum distance-based separation rules for particular aircraft landings sequences under IMC (Instrument Meteorological Conditions) have been specified. As such, they have secured safe operations by preventing impact with the wake vortices behind particular categories of landing aircraft on the one hand, and elimination of distance "buffers" introduced to compensate the above-mentioned variations of aircraft approach speeds on the other. Consequently, they have contributed to increasing the runway landing capacity under specified conditions [9]. In addition, if it was possible for pilots to more precisely monitor the behaviour of the wake vortices behind particular aircraft categories, ATC controllers, and/or both, the time-based separation rules could be based on such behaviour, which would bring them closer in line with those based on existing ATC distance-based separation rules applied to VMC (Visual Meteorological Conditions) in the United States. These rules would be shorter than those applied under IMC, which would consequently increase the corresponding runway landing capacity without compromising safety due to potential impact with wake vortex. In addition, they could be applied independently of weather conditions, which would make such higher runway landing capacity fairly stable and independent from weather changes. As mentioned above, this may be of particular value to US airports practicing landings under both VMC and IMC. The two parameters—ceiling and visibility—determine the boundary conditions between IMC and VMC as shown in Fig. 5.1 for selected US airports [10].

As can be seen, the critical ceiling is most diverse when the horizontal visibility is 3 and 5 (statute) miles and relatively homogeneous when this visibility is 4, 7, and 8 miles. In addition, most of these airports operate within the margin between "high IFR" and "marginal VFR" [10, 21]. Depending on the weather conditions (VMC or IMC), the ATC applies VFR and IFR distance-based separation rules between landing aircraft given in Table 5.2.

As can be seen, current IFR separation rules are around 40% stricter than VFR distance-based separation rules. Such differences cause differences in the corresponding landing capacities and consequently create an inherent sensitivity of these capacities to weather changes. Both separation rules generally eliminate the impact of the wake vortices of leading aircraft in particular combinations of

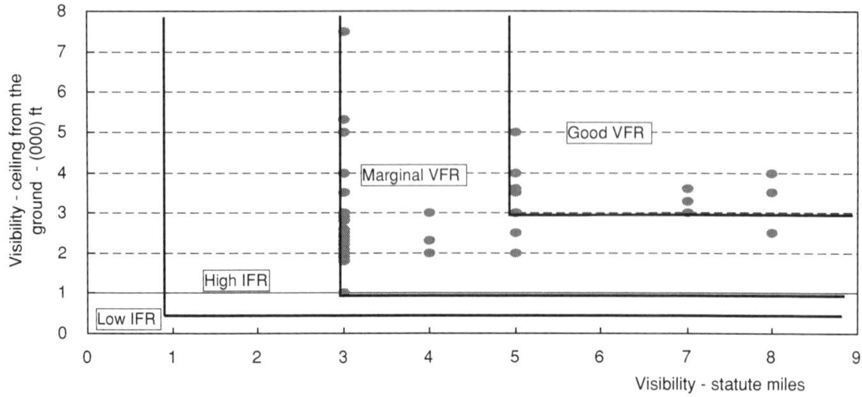

Fig. 5.1 Characteristics of meteorological boundary conditions at 75 selected US airports (Compiled from [10, 21])

Table 5.2 FAA/ICAO separation rules for landing aircraft (nm)

i/j	Small	Large	B757	Heavy
VFR				
Small	1.9	1.9	1.9	1.9
Large	2.7	1.9	1.9	1.9
B757	3.5	3.0	3.0	2.7
Heavy	4.5	3.6	3.6	2.7
IFR				
Small	2.5(3)	2.5(3)	2.5(3)	2.5(3)
Large	4.0	2.5(3)	2.5(3)	2.5(3)
B757	5.0	4.0	4.0	4.0
Heavy	6.0	5.0	4.0	4.0

Compiled from [10, 20, 21]

landing sequences. Under the assumption that the potential exposure of trailing aircraft to the wakes generated by leading aircraft is virtually the same for both types of separation rules, the question of the difference between VFR and IFR separation rules arises. The possible answer could be that under VMC, trailing aircraft fly on the principle "see and be seen" with just sufficient safe distance to avoid the wake vortex hazard from the leading aircraft. Under IMC, in addition to basic separation to avoid the wake vortices, ATC introduces additional "buffers" to compensate the cumulative system error in estimating the aircraft position(s) visualised for ATC controllers thanks to the sophisticated radar system. In addition, these "buffers" compensate the above-mentioned deviations in landing speeds due to wind. In Europe, independent of the weather conditions, landings are carried out exclusively according to the IFR separation rules in Table 5.1. Figure 5.2a, b shows the effect on the landing capacity at selected US and European airports.

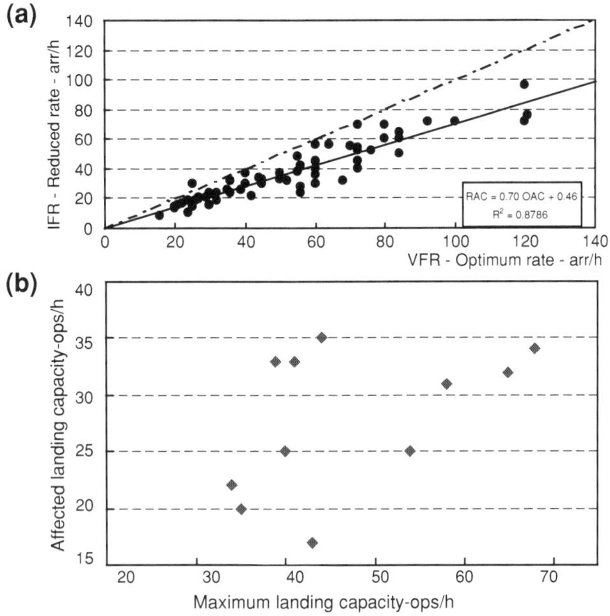

Fig. 5.2 Landing capacity affected by weather conditions (Compiled from [10, 21]). **a** US airports. **b** European airports

Figure 5.2a shows that at most US airports, there is an average difference of about 30% between IFR and VFR landing capacities. Due to their rather strong linear relationship, IFR landing capacities generally amount up to about 70 % of the corresponding VFR landing capacities. Figure 5.2b shows that at major European airports, bad weather significantly affects the declared landing capacity despite the exclusive use of IFR separation rules. The main reasons are increased ATC separation rules in poor weather on the one hand and temporary closure of runways on the other. Consequently, the suggested time-based separation rules will need to be standardised respecting the true (dynamic) behaviour of the wake vortices and thus provide more stable landing capacities under IMC comparable situation or even greater capacities than those under VMC.

5.3.1.2 Configuration of the "Wake Reference Airspace"

The wake vortices behaviour is monitored dynamically in the "wake reference airspace" used for the final approach and landing on a given runway. In general, this airspace consists of two parts: (1) the "wake vortex corridor", i.e. the prismatic airspace which spreads along the extended centreline of the runway; and (2) the SHA (Simplified Hazard Area) in which the wake vortex generated by a given aircraft remains until decaying and/or vacating the "wake reference airspace" [16]. Figure 5.3 shows a simplified three-dimensional diagram of this airspace where: γ is the length of the "wake vortex corridor"; Δ is the horizontal

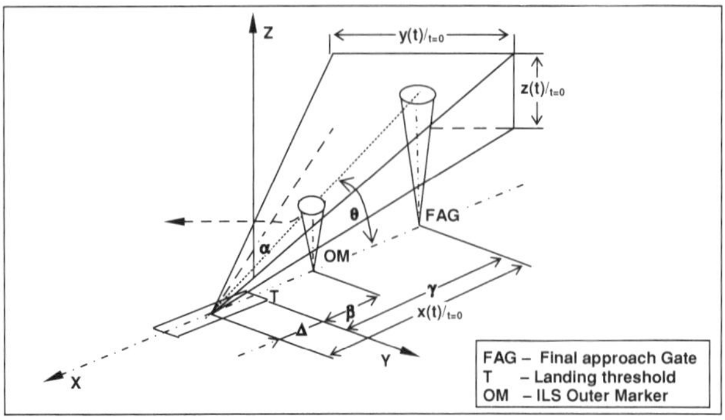

Fig. 5.3 Three-dimensional diagram of the "wake reference airspace" (Compiled from [16])

distance between FAG at the beginning of the "wake vortex corridor", and the runway landing threshold *T*; OM and MM are Outer and Middle marker, respectively, of the ILS (Instrument Landing System); $x(t)$, $y(t)$, and $z(t)$ are longitudinal, horizontal, and vertical coordinates, respectively, of the "wake reference airspace", depending on time (t); β is the horizontal distance between the location of OM and the runway landing threshold *T*; is the angle between the axis of the "wake vortex airspace" and one of its sides in the horizontal plane; and θ is the nominal angle (ILS Glide Slope) of the aircraft approach path in the "wake vortex airspace".

Monitoring and prediction of wake vortex behaviour in the "wake reference airspace" is and will be carried out by current and forthcoming technologies and systems both on the ground and on board the aircraft (see Table 5.1) [3]. The most well-known current system used on the ground is AVOSS (Aircraft Vortex Spacing System) currently operating at Dallas Fort Worth airport (US). The system provides the dynamic spacing criteria between aircraft approaching the single runway along a pre-defined corridor. These criteria are based on the prediction of the wake vortex position and strength dependent on the current weather conditions. The wake attribute, which first clears the corridor at a certain ("reference") profile, defines the distance separation criterion for a given aircraft. More advanced systems such as ATC WAKE, WAKEVAS, and WVDS will be able to provide even more accurate information including its exchange between pilots and ATC controllers online, i.e. automatically via data link [3, 9]. This information indicates the strength and prospective behaviour (movement) of the wake vortex within the "wake reference airspace and is presented on the cockpit screen (the Navigational or Primary Flight Display). This will allow pilots to monitor the wake vortex of the aircraft they follow instead of looking at the aircraft itself which they can see under VMC and not under IMC. Consequently, separation between landing aircraft will become purely based on dynamic time-based separation rules, which will, in terms of separation distances, come close to today's VFR separation distances.

5.3.2 Prioritising Aircraft Landings

5.3.2.1 Background

Airport/airline scheduling usually matches airport demand in terms of the number of landings and take-offs, and the corresponding strategic runway system capacities. The outcome is the airport's "practical" capacity as the trade-off between the specified level of utilisation of the airport runway "ultimate" capacity and the level of acceptable average delay per landing and/or take-off under the specified (congestion) conditions. This trade-off is agreed between the airport, local ATC, and the airlines. Currently, airline scheduling practice dictates the order of serving particular aircraft categories (classes) which is generally respected by the ATC as the First-Come–First-Served (FCFS) priority rule, i.e. according to the times of requests for service. Any eventual exception from this rule in terms of prioritising particular categories of aircraft/flights could be considered for increasing the runway's "ultimate" capacity and as an additional alternative to introducing time-based separation rules [18].

5.3.2.2 Conditions for Implementation

In order to eventually introduce prioritising of aircraft landings and/or take-offs, a variety of market-policy and tactical/operational conditions need to be simultaneously fulfilled.

The *Market-policy conditions* are as follows:

- An airline and/or its alliance should have a predominant number of slots at the given airport during the congestion period. This would diminish eventual discrimination between airlines since it is assumed that a given airline and/or its alliance partners have already self-internalised their total delays and related congestion costs. In reality, this condition is fulfilled at many airports. For example, there is reason to believe that airline alliances ultimately dominate their European hubs. The STAR alliance shares about 70% and its leading airline Lufthansa 59% of slots at their primary hub Frankfurt Main airport (Germany). The SkyTeam alliance and its leading airline Air France-KLM share 63 and 57% of slots, respectively, at their main hub Paris Charles de Gaulle airport (France). The Oneworld alliance and its leading airline British Airways share about 20 and 41% of slots, respectively, at their primary hub London Heathrow airport (UK) [6, 7]. In addition, at 30 of the busiest US airports, the average share of the first dominant airline in 2007 was about 43% (standard deviation 19%), while the average share of the second dominant airline was 17% (standard deviation 7%) [23].
- The aircraft fleet using the same runway(s) should be heterogeneous in terms of the aircraft wake vortex categories, implying their distinctive size (number of seats, MTOW (Maximum Take-Off Weight)), approach speeds, operating cost,

and consequently the unit cost of delays. For example, at the above-mentioned large European airports, the aircraft fleet mix during the peaks, which coincides with the banks of incoming and outgoing mutually connected flights, consists of about 15–20% heavy, 70–75% large, and 10% small aircraft, thus making it sensible to prioritise them during landing and/or take-off on the same runway(s). If delay cost is a criterion for prioritising, larger aircraft should have priority, contributing to increasing the aircraft size by deterring access of smaller aircraft, and consequently ignoring particularly thin routes/markets. Alternative criteria could include assigning priorities to the more conveniently equipped aircraft and/or to specific aircraft/flights of particular airlines. The latter implies that each airline would be assigned a number of slots during particular periods of time (for example, one hour) with absolute priority enabling delay-free flights.

The *tactical/operational conditions* are as follows:

- It should be possible to assign priorities under both IMC (IFR) and VMC (VFR). In both cases, ATC should be fully responsible for allocating priorities to particular atm categories in addition to providing assistance in monitoring their separation;
- Assigning priorities should be carried out exclusively at the tactical/operational level. This implies planning by the local ATC (ATM), clustering particular atms into homogeneous clusters, and serving particular clusters according to the pre-assigned priorities during the period, which can vary from a few hours (tactical) to an hour or half an hour (operational);
- The atms of particular priority classes should be assigned dedicated TMA entry/exit gates, and consequently dedicated arrival/departure trajectories. If different clusters of arriving atms are assigned the same TMA entry point, those with higher priority should always be assigned lower entry altitudes. Similarly, departing atms should be clustered while still on the apron parking gates/stands.
- The assignment of particular TMA entry/exit gates should be carried out according to different criteria. For example, one criterion could include allocating the entry gate closer to the FAG (Final Approach Gate) of the landing runway to larger aircraft, enabling such aircraft the shortest approach trajectories through TMA. In addition, the same cluster of atms could be assigned more than one TMA entry gate. When a holding pattern is needed, which usually happens, additional airspace will be needed, which can be a problem at some already congested airports.
- Prioritising atms should not impose additional delays and consequently compromise airline schedule(s). Once the procedure is agreed, airlines could incorporate prospective delays into their schedules. This would appear to be particularly relevant for delays imposed on lower priority atms, which should at least be predictable and controllable; and
- Prioritising the same categories of atms should be carried out according to the FCFS priority rule.

Operationalization

This implies developing an algorithm as a component of a system such as the above-mentioned CTAS (Centre/TRACON Automated System), integrated Arrival/Departure Manager, DA (Descent Advisor), and FAST (Final Approach Spacing Tool) [24]. In any case, this algorithm should take into account the following:

- Since new (prioritising) practices depart from the currently used FCFS practice, they should not create additional ATC controllers' workload, but rather minimise it if possible. In the case at hand, the effect on ATC controllers' workload appears to be neutral since clustering and assigning dedicated TMA entry gates and related arrival trajectories to the different atm clusters eliminates the potential overtaking conflicts in the arrival sequence "slow"–"fast". In addition, pairing arriving aircraft from different clusters at FAG becomes simpler: namely, the clusters are processed at FAG of the same runway after each other. Thus, the last aircraft from the higher priority cluster should only be paired with the first aircraft from the lower priority cluster, and vice versa;
- The algorithm should be applied only if the atms from particular clusters have the same or nearly the same Estimated Arrival Time (ETA) at the FAG of the given runway. This implies that demand certainly exceeds the "ultimate" runway capacity; and
- If only savings of total delays are of interest, the order of serving particular atm clusters is not relevant. However, if the total cost of delays is the main criterion, this order becomes crucially relevant [5, 18].

5.3.3 The ATC Vertical Distance-Based Separation Rules

5.3.3.1 Background

Advanced operational procedures for single-runway approach, which could be convenient for application of the ATC vertical distance-based separation rules, can be classified into three categories: (1) single segment with constrained GS angles; (2) double segment with constrained GS angles; and (3) single segment with ultimately unconstrained GS angles.

Single Segment with Constrained GS Angles

This final approach procedure is supported by MLS or the multiple ILS GP (Instrumental Landing System Glide Path), which enables particular categories of aircraft to use different GS angles along the entire final approach path—from the FAG (Final Approach Gate) to landing threshold T (see Fig. 5.3). In such cases, certified aircraft use a steeper GS angle (>3°) while other aircraft use the nominal ILS GS angle (3°). ATC applies horizontal distance-based separation rules to landing sequences in which all aircraft use the nominal ILS GS angle, and vertical separation rules in landing sequences in which one aircraft (usually the leading one) uses the nominal and other aircraft (usually those trailing) use the steeper GS angle. Regardless of the aircraft sequence, the minimum ATC vertical distance-based separation rule is usually 1,000 ft. In any case, at least one type of separation rule between the aircraft in particular sequences must be satisfied.

Double Segment with Constrained GS Angles

This final approach procedure is based on the concept of "Individual Flight Corridor(s)" (i.e. 4D RNAV trajectories connecting WP (Way Point) starting at the intermediate approach, the FAG, and the landing threshold). Originally, this concept had been designed for landing on closely spaced parallel runways. 4D final approach trajectories consist of an Outer and Inner segment. As applied to a single runway, these 4D trajectories overlap in the horizontal and vertical plane along the Inner segment. They can overlap and/or differ in both horizontal and vertical planes along the Outer segment. This implies that the Inner segment has the common GS angle for all aircraft categories. The Outer segment can have different and/or the same (flexible) GS angles for particular aircraft categories. Thus, the prime distinction from the case (1) is that all aircraft use the common nominal GS angle (3°) along the Inner segment and different GS angles (3–6°) along the Outer segment. Under such circumstances, a mixture of ATC horizontal and vertical distance-based separation rules can be applied to the aircraft along the Outer segment, while horizontal distance-based separation rules are exclusively applied to aircraft in the Inner segment. Consequently, thanks to the vertical distance-based separation rules, the aircraft in particular sequences can come safely closer to each other while always maintaining at least one type of ATC minimum separation rules.

Single Segment with Ultimately Unconstrained GS Angles

This is the rather radical final approach procedure based on 4D RNAV trajectories consisting of a single straight and/or curved segment connecting FAG and the runway landing threshold. For different aircraft of the same and/or different categories, this segment can differ in the horizontal plane, except for the short common portion just in the vicinity of the runway landing threshold. This segment is used for setting up the full landing configuration and speed. In the vertical plane, the entire trajectory is different for particular aircraft categories due to different GS angles assigned. In addition, each of these trajectories can be considered as the final part of a CDA (Continuous Descent Approach) trajectory aimed at reducing flight time, fuel consumption and related emissions of greenhouse gases, and noise around particular airports. Thus, the landing aircraft can be assigned different GS angles (3–6°). Consequently, contrary to cases (1) and (2), the trajectories of the leading aircraft in all sequences can be higher, equal to, and/or lower than those of the trailing aircraft, and vice versa. In addition, as in case (2), the 4D RNAV trajectories are supported by multiple ILS GP or MLS, as well as other above-mentioned innovative new technologies.

The diversity of rather small GS angles of 4D RNAV final approach trajectories offers ATC an opportunity to exclusively apply vertical distance-based separation rules to all sequences of landing aircraft. In such a way, depending on the location where this minimum vertical separation is established (the FAG or the runway landing threshold), the trailing aircraft can in almost all sequences come closer to the leading aircraft while always being above and thus fully protected from the wake vortices which usually move behind and below the flight paths of the leading aircraft. Consequently, the horizontal distances between particular aircraft become generally shorter than in cases (1) and (2), when exclusively horizontal distance or a mixture of the horizontal and vertical distance-based separation rules are applied, respectively.

5.3.3.2 Conditions for Implementation

In order to eventually implement the above-mentioned new operational procedures (2) and (3) and eventually apply either a mixture of horizontal and vertical-distance based separation rules, or exclusively vertical distance-based separation rules, three sets of conditions need to be fulfilled: supportive technologies and decision support tools, aircraft certification for steeper GS angles, and training of both ATC controllers and pilots.

New technologies in Table 5.1 supporting the above-mentioned procedures need to be implemented.

Since GS angles greater than (4–4.5°) are considered steeper, aircraft need to be certified for such angles. In the given context, certification for a given range of GS angles needs to become the rule rather than the exception as it is at present and include virtually all small, the majority of large, and B757 aircraft (Table 5.2) in order to make the eventual application of the vertical distance-based separation rules sensible. In principle, smaller aircraft can be certified for steeper GS angles than larger aircraft. What is the real change of approaching at a steeper GS angle? If DH (Decision Height) above the landing threshold remains 50 ft (15 m) and the flare distance 35 ft (10.5 m), the change of GS angle from 3 to 5.5° will shorten the flare and touchdown distance from 300 and 160 m to 200 and 110 m, respectively [6].

Pilots need to be trained for such steeper approaches in order to provide safe sustainable operations in the long term. In addition, if CDTI is used, some changes in order to improve awareness about the vertical separation based on the horizontal distances shown on the cockpit's screen will be needed.

The ATC controllers will need training, which will help in upgrading their current mental models and the way of sequencing landing aircraft. In particular, a convenient tool for converting horizontal distances shown on the radar screed into vertical separation rules to be maintained will be needed, albeit not compulsorily.

5.4 A Methodology for Estimating Effects of Advanced Technologies, Procedures, and Regulations on Runway Landing Capacity

5.4.1 Background

The methodology for estimating the effects of the above-mentioned advanced operational regulations, procedures, and supporting technologies consists of three dedicated models for estimating the runway ultimate landing capacity when: (1) ATC time-based separation rules are applied; (2) landings are prioritized; and (3) ATC vertical distance-based separation rules are applied. With regard to their structure, these analytical models enable a sensitivity analysis of the runway landing capacity to be carried out with respect to changes to the most influencing factors. They are applied to the generic case of an airport landing runway using the "what-if" scenario approach.

5.4.1.1 Previous Research

Analysis and modelling of ultimate airport (runway) capacity have occupied airport officials, ATC, airlines, planners, analysts, and academics for years. These efforts have resulted in developing numerous analytical and simulation models, which could be classified into two broad classes for: (1) calculating the (runway) capacity of individual airports and the capacity of airport network(s) [22]; and (2) optimization of utilization of the airport (runway) capacity under changing influencing factors [1, 28].

Specifically, analytical models have provided two-value parameters—one for the arrival and another for the departure capacity [2, 12–14, 19, 27].

In addition, an analytical model of the runway landing capacity based on partial use of ATC vertical distance-based separation rules, possible due to steeper approach and landing of particular aircraft types thanks to MLS, has been developed. This concept has been recently extended to modelling the ultimate landing capacity of closely spaced parallel runways [17]. In addition, models for estimating the ultimate landing capacity of a single runway based on ATC time-based separation rules and prioritizing aircraft operations have been developed all as part of the current NextGen (US) and SESAR (Europe) programs (http://vams.arc.nasa.gov/activities/tacec.html, [16, 18]).

Some other models such as the FAA Airport Capacity Model, LMI Runway Capacity Model, and DELAYS as 'Quasi-Analytical Models of Airport Capacity and Delay', developed mainly for airport (runway) planning purposes and based on the analytical single-runway capacity model, have calculated the so-called "capacity coverage curve" including the associated aircraft delays [12, 19]. At the same time, separate models of the ultimate capacity of the apron/gate complex and the system of taxiways have been developed. Only recently, efforts have been made to integrate these analytical models into an 'airport integrated-strategic planning tool'. Such integration has however been achieved by developing computer-supported simulation models for calculating the airport capacity and delay at (1) Low (HERMES and The Airport Machine), (2) Intermediate (NASPAC, FLOWSIM and TMAC), and (3) High Level of Detail (TAAM and SIMMOD) [15, 22, 27]. In comparison to the analytical models, these models have studied the airport airside operations in much greater detail [22].

5.4.1.2 Objectives and Assumptions

The objective is to develop a methodology for estimating potential gains in runway "ultimate" landing capacity which could be obtained through the above-mentioned advanced procedures and regulations supported by advanced technologies. In addition, this methodology must enable sensitivity analysis of landing capacity with respect to the most important influencing factors. The methodology, consisting of three models, is based on the following assumptions, [16–18]:

- The three-dimensional approach and landing trajectories of particular aircraft categories in the "wake vortex reference space" are known beforehand; assignment of conventional and/or steeper approach trajectories depends on the type of arrival sequence(s) in terms of the aircraft wake-vortex category, approach speeds, and certified capability to perform either approach safely;
- Aircraft change their approach speeds at the particular locations in the "wake reference airspace" almost instantly;
- Due to advanced technologies, monitoring of the current and prediction of the prospective behaviour of wake vortices in the "wake reference airspace" is reliable;
- The influence of weather on wake vortex behaviour for a given landing sequence is constant during the aircraft's occupation of "wake reference airspace";
- Aircraft appear on the particular segments of approach trajectories at the exact moment the ATC expects them to; this implies that the time and space deviations of the actual from the prescribed aircraft positions in both horizontal and vertical

planes are considered negligibly small, mainly due to the above-mentioned new technologies which are assumed to be in place at the time (Table 5.1);

- The appearance of particular aircraft categories in particular landing sequences are mutually independent events;
- The runway landing threshold is the "reference location" for determining "capacity";
- ATC uses either a mixture of radar-based horizontal and vertical, or exclusively vertical distance-based separation rules between landing aircraft; alternatively, it may use the time-based separation rules; and
- Consensus exists between ATC, airlines, and the airport on prioritising particular arriving aircraft instead of using the common FCFS rule.

5.4.2 The Basic Structure of the Models

The models constituting the above-mentioned methodology possess a common basic structure, which implies determining the "ultimate" landing capacity of a given runway as the reciprocal of the minimum average "inter-arrival" time of passing of all combinations of aircraft pairs through a given "reference location" selected for their counting during a given period of time under conditions of constant demand for service [2]. In the given context, the "reference location" for counting aircraft operations is the runway landing threshold. The minimum average inter-arrival time enables maximum aircraft operations at the landing threshold and consequently flow at full capacity. The period of time is a quarter, half, and/or an hour. Under such conditions, the basic structure of the model when FCFS (First-Come–First-Served) service discipline is applied to aircraft landing on a single runway is based on the traditional analytical model of runway "ultimate" landing capacity as follows [2, 15]:

$$\lambda_{\mathrm{a}} = T / \sum_{\mathrm{ij}} p_{i\,\mathrm{a}} t_{ij\mathrm{min}}\, p_j \tag{5.1}$$

where

$_{\mathrm{a}}t_{ij\mathrm{min}}$ is the minimum inter-arrival time of the aircraft pair (i) and (j) at the runway landing threshold selected as the "reference location" for counting operations;

p_i, p_j is the proportion of aircraft types (i) and (j) in the landing mix, respectively;

T is the period of time (usually one hour).

In the case of introducing priorities (PR) instead of the conventional FCFS rule, expression (5.1) transforms as follows [18]:

$$\lambda_a = T / \sum_{i=1}^{N} p_i t_{ii/\min} \tag{5.2}$$

where

$t_{ii/\min}$ is the minimum inter-arrival time between a pair of aircraft of priority class (*i*) at the "reference location" (*i* – 1, 2, N);

$p_{i,}$ is the proportion of aircraft of priority class (*i*) in the landing mix.

Expression (5.2) implies that the arriving aircraft are clustered into groups consisting of the same aircraft priority classes (categories), which are then served sequentially one after the other. The order of their serving does not influence the average service rate, i.e. runway "ultimate" landing capacity.

5.4.3 Determining the Aircraft Inter-Arrival Time(s) at the "Reference Location"

5.4.3.1 The ATC Time-Based Separation Rules

Dealing with ATC time-based separation rules for aircraft landing on a single runway includes modelling the wake-vortex characteristics and behaviour in the "wake reference airspace", setting up dynamic time-based rules, and calculating the inter-arrival times of particular sequences of landing aircraft at the runway threshold *T* in Fig. 5.3 [16].

Wake Vortex Characteristics and Behaviour

The wake vortex appears as soon as lift is created on the aircraft wings. Investigation so far has shown that wakes behind aircraft decay over time generally at a disproportional rate, simultaneously descending below the aircraft trajectory at a certain descent speed. Without crosswind they move from the aircraft trajectory at a self-induced speed of about 5 kt (knots). Otherwise, they move according to the direction and speed of crosswind [26].

Modelling wake-vortex behaviour includes determining its strength, i.e. root circulation, the "reference time", decaying pattern, descent speed, and movement influenced by ambient weather.

The wake strength the root circulation at time (*t*)

This can be estimated as follows:

$$\Gamma_0(t) = \frac{4Mg}{\rho v(t) B\pi} \tag{5.3a}$$

The wake reference time-the time in which the wake descends one wing span at time (t)

This is estimated as follows:

$$t^{*}(t) = \frac{\pi^{3}B^{2}}{8\Gamma_{0}(t)} = \frac{\rho\pi^{4}B^{3}v(t)}{32Mg} \tag{5.3b}$$

The wake-decaying pattern

This is estimated as follows:

$$\Gamma(t) = \Gamma_{0}(t)\left(1 - \frac{t}{kt^{*}(t)}\right) \tag{5.3c}$$

If the safe wake strength is Γ*, the time the wake needs to decay to this level, τ_d (Γ*) can be determined from expression (5.5c) as follows:

$$\tau_{d}(t, \Gamma^{*}) = kt^{*}(t)\left(1 - \frac{\Gamma^{*}(t)}{\Gamma_{0}(t)}\right) \tag{5.3d}$$

The wake's self-induced descent speed

This is determined as follows:

$$w(t) = \frac{2\Gamma(t)}{\pi^{2}B} = \frac{2\Gamma_{0}(t)[1 - t/kt^{*}(t)]}{\pi^{2}B} \tag{5.3e}$$

where

- M is the aircraft (landing) mass (kg);
- g is gravitational acceleration (m/s^2);
- ρ is air density near the ground (kg/m^3);
- $v(t)$ is aircraft speed at time (t) (m/s);
- B is the aircraft wingspan (m); and
- k is the number of reference time periods after the wakes decay to the level of natural turbulence near the ground (70 m^2/s) ($k = 8$–9).

The ambient weather characterised by ambient wind can influence the wake vortex behaviour in the "wake reference airspace". This wind is characterised by two components: crosswind and headwind.

Crosswind can be determined as follows:

$$V_{cw}(t) = V_{w}(t)\sin(\varphi_{w} - \varphi_{a}) \tag{5.3f}$$

Headwind can be determined as follows:

$$V_{hw}(t) = V_{w}(t)\cos(\varphi_{w} - \varphi_{a}) \tag{5.3g}$$

where

- $V_w(t)$ is the wind reported by ATC at time (t);
- φ_w is the course of the wind;
- φ_a is the course of the aircraft.

The wake vacates the "reference profile" at almost the same speed as the crosswind.

The headwind does not directly influence the wake descent speed (rate) but does move the wake from the ILS GS and thus increases its vertical distance from the trailing aircraft's path. This vertical distance increases linearly over time and is proportional to headwind as follows:

$$\Delta z_{\mathrm{hw}}(t) = V_{\mathrm{hw}}(t) \times t \times tg\theta \tag{5.4h}$$

where all symbols are as in the previous expressions.

Dynamic Time-Based Separation Rules

Let $\tau_{ij/\min}(t)$ be the minimum time-based separation rule between leading aircraft (i) and $t(j)$ in landing sequence (ij) at time (t). Currently, this time depends on ATC distance-based separation rules (either IFR or VFR) implicitly including the characteristics of wake vortex behaviour and aircraft approach speeds (see Table 5.4). The main idea is to make these time separations explicitly based on the current and predicted characteristics and behaviour of the wake vortex generated by the leading aircraft (i) in the given sequence (ij). The characteristics and behaviour of the wake vortex include its initial strength and time of decay to a reasonable (i.e. safe) level, and/or the time of clearing the given profile of the "wake reference airspace" either by self-induced descent speed, headwind, self-induced lateral speed, and/or crosswind.

Let $\tau_{ij}(t)$, $\tau_{iy}(t)$ and $\tau_{iz}(t)$, respectively, be the time separation intervals between aircraft (i) and (j) based on the current ATC distance-based separation rules in Table 5.1, and the predicted time it takes for the wakes of leading aircraft (i) to move either horizontally or vertically at time (t) out of the "wake reference profile". In addition, let $\tau_{id/j}(t)$ be the predicted time of decay of the wake of the leading aircraft (i) to the level acceptable for the trailing aircraft (j) at time (t). Referring to Fig. 5.3, these times can be estimated as follows:

$$\begin{aligned} &\tau_{ij}(t) = \delta_{ij}(t)/v(t) \\ &\tau_{iy}(t) = Y_i(t)/V_{\mathrm{cw}}(t) \\ &\tau_{iz}(t) = \min[Z_i(t)/w_i(t); \Delta z_{ij/\min}(t)/V_{\mathrm{hw}}(t)tg\theta] \\ &\tau_{id/j}(t,\Gamma^*) = kt_i^*(t)\left[1 - \Gamma_j^*/\Gamma_{0i}(t)\right] \end{aligned} \tag{5.5a}$$

where

$\delta_{ij}(t)$ is the minimum ATC horizontal distance-based separation rule applied to landing sequence (ij) at time (t);

$v_j(t)$ is the average approach speed of the trailing aircraft (j) at time (t); and

$\Delta z_{ij/\min}(t)$ is the minimum vertical separation rule between aircraft (i) and (j) at time (t).

Other symbols are analogous to those in previous expressions. Expression (5.7a) indicates that the time the wakes take to move out of the "reference profile" does not depend on the type of trailing aircraft. However, the decay time of the wake vortex generated by the leading aircraft depends on its strength, which has to be acceptable (i.e. safe) for the trailing aircraft. Consequently, at time (t), the trailing aircraft (j) can be separated from the leading aircraft (i) by the following minimum time separation rule:

$$\tau_{ij/\min}(t) = \min\left[\tau_{ij}(t); \tau_{iy}(t); \tau_{iz}(t); \tau_{id/j}(t, \Gamma^*)\right] \tag{5.5b}$$

- If $v_i \leq v_j$, the minimum time separation rule $\tau_{ij/\min}$ (t) needs to be established when the leading aircraft (i) is at the runway landing threshold T in Fig. 5.3, i.e. at time $t = \gamma/v_i$. In addition, the following condition must be fulfilled: $\tau_{ij/\min}$ (t) $\geq t_{ai}$, where t_{ai} is the runway occupancy time of the leading aircraft (i).
- If $v_i > v_j$, the minimum time separation rule $\tau_{ij/\min}$ (t) should be established when the leading aircraft (i) is just at Final Approach Gate (FAG) in Fig. 5.3, i.e. at time $t = 0$. This is based on the fact that the faster leading aircraft (i) will continuously increase its distance from the slower trailing aircraft (j) during its runway approach.

Minimum Inter-Arrival Times Between Successive Landings

The minimum inter-arrival times for aircraft sequences (i) and (j) at the landing threshold can be determined based on expression (5.5b) as follows:

$$_{a}t_{ij/\min} = \left\{ \begin{array}{ll} \tau_{ij/\min}(t = 0) + \gamma(1/v_j - 1/v_i) & \text{for } v_i > v_j \\ \max\left[t_{ai};\ \tau_{ij/\min}(t = \gamma/v_i)\right] & \text{for } v_i \leq v_j \end{array} \right\} \tag{5.5c}$$

where $\tau_{ij/\min}(t)$ is determined according to expression 5.5a and 5.5b.

At time $t = 0$, when the leading aircraft (i) is at the FAG, the "wake reference profile" is at its greatest, which implies that the wakes need the longest time to vacate it by any means. At time $t = \gamma_i/v_i$, when the leading aircraft (i) is at the landing threshold, the "wake reference profile" is the smallest, which implies that the wakes need a shorter time to vacate it (see Fig. 5.3).

5.4.3.2 Prioritising Aircraft Landings

The minimum inter-arrival times between the sequences of aircraft on the landing threshold T, $t_{ii/\min}$ in expressions (5.1) and (5.2) for FCFS and PR (Priority) rule, respectively, can be estimated as follows [18]:

(a) For the FCFS rule:

$$t_{ij/\min} = \left\{ \begin{array}{ll} \delta_{ij}/v_j; & \text{for } v_i \leq v_j \\ \delta_{ij}/v_j; + \gamma(1/v_j - 1/v_i); & \text{for } v_i > v_j. \end{array} \right\} \tag{5.6a}$$

(b) For the priority (PR) rule:

$$t_{ii/\min} = \delta_{ii}/v_i \tag{5.6b}$$

where the symbols are analogous to those in the previous expressions. Expression (5.8a) is given for comparison. Expression (5.6b) applies to all sequences in the priority cluster (*i*) of aircraft landing on a given runway.

5.4.3.3 The ATC Mixed and Vertical Distance-Based Separation Rules

The Mixed Horizontal/Vertical Distance-Based Separation Rules

A mixture of ATC minimum horizontal and vertical distance-based separation rules can be applied between particular sequences of landing aircraft using the above-mentioned Individual Flight Corridors with two segments, each with different GS angles. In such cases, the minimum inter-arrival time of aircraft in sequence (*ij*) can be estimated as follows:

$$t_{ij/m} = \begin{bmatrix} \delta_{ij}/v_j & \text{for } v_i < v_j \\ H^0_{ij}/v_j \sin\theta_j, & \text{for } v_i = v_j \\ H^0_{ij}/v_j \sin\theta_j + \gamma_{ij}(1/v_j - 1/v_i), & \text{for } v_i > v_j \end{bmatrix} \tag{5.7a}$$

where

$H^0_{ij}j$ is the minimum ATC vertical distance-based separation rule between the aircraft of wake vortex categories (*i*) and (*j*);

θ_j is the GS angle of aircraft (*j*), which can be different for the Outer and Inner segments of the final approach trajectory.

The other symbols are analogous to those in expression (5.3a).

Expression (5.7a) indicates that ATC vertical separation rules are applied to the aircraft sequence of the same speed at the landing threshold and to the "fast–slow" sequence at the FAG. The ATC horizontal separation rules are applied to the "slow–fast" aircraft sequence. Under such conditions, at least one separation rule is guaranteed to ensure safe landing of particular aircraft sequences.

Vertical Distance-Based Separation Rules

ATC minimum vertical distance-based separation rules can be exclusively applied to the particular landing sequences in which the aircraft use different GS angles along the entire final approach trajectory. Depending on approach speeds and GS angles, there can be twelve different combinations of landing sequences, for which the minimum inter-arrival times at the reference location, i.e. landing threshold, is calculated as follows:

$$t_{ij/v} = \begin{bmatrix} H_{ij}^{0}/v_j \sin\theta_j, & \text{for } v_i \le v_j \text{ and } v_i \sin\theta_i >=< v_j \sin\theta \\ & \text{for } v_i > v_j \text{ and } v_i \sin\theta_i < v_j \sin\theta_j \\ H_{ij}^{0}/v_j \sin\theta_j + \gamma_i tg\theta_i(1/v_j \sin\theta_j - 1/v_i \sin\theta_i) & \text{for } v_i > v_j \text{ and } v_i \sin\theta_i >= v_j \sin\theta_j \end{bmatrix} \tag{5.7b}$$

where

γ_i is the length of the final approach path of aircraft (*i*);

θ_i is the GS angle of the final approach path of aircraft (*i*).

The other symbols are analogous to those in the previous expressions.

In the first term of expression (5.7b), the first condition indicates that horizontal separation remains constant or decreases while vertical separation remains constant, decreases, or increases. The second condition indicates that horizontal separation increases while vertical separation decreases. Under such circumstances, in order to minimise the inter-arrival time $t_{ij/v}$, ATC minimum vertical distance-based separation rules should be established at the moment when the leading aircraft (*i*) in the sequence (*ij*) is just at the landing threshold *T* in Fig. 5.1.

In the second term of expression (5.7b), the condition indicates that the horizontal separation between aircraft (*i*) and (*j*) increases while the vertical separation remains constant or also increases. Under such conditions, in order to enable the minimum inter-arrival time $t_{ij/v}$ at the landing threshold, ATC minimum vertical separation rules should be established at the moment when the leading aircraft (*i*) in sequence (*ij*) is at the FAG. This case corresponds to aircraft sequence (*kl*) in Fig. 5.4.

In expressions (5.5c), (5.6a), (5.6b), (5.7a), and (5.7b), the minimum inter-arrival time ${}_{a}t_{ij\min}$ must be at least equal to or greater than the runway landing occupancy time t_{ai} of the leading aircraft (*i*) in the sequence (*ij*). This ensures that only one aircraft occupies the runway at any time.

5.4.4 Criteria for Selecting ATC Separation Rules

Expressions (5.3a)–(5.7b) suggest that ATC may be able to select and apply any combination of horizontal distance or time-based and vertical distance-based separation rules to the particular sequences of landing aircraft. This will enable truly minimal inter-arrival times at the runway landing threshold, and consequently maximise the landing capacity under given conditions while at the same time fully satisfying the safety requirements.

In general, this process can be carried out as follows.

Given the characteristics of the aircraft flow expected to land on a given runway over the forthcoming specified period of time, the priority rule—FCFS or PR—can be chosen. The control function **Z** with only two binary values—0 or 1, the former for FCFS service discipline and the latter for PR, can be used.

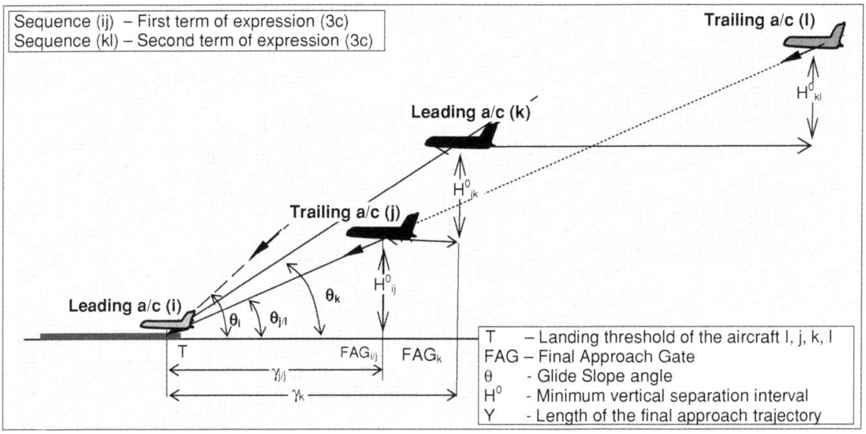

Fig. 5.4 Illustration of some characteristic cases of applying ATC vertical distance-based separation rules between landing aircraft (Referring to expression (5.7b))

Then, the separation rule to be applied between particular aircraft sequences can be selected. For sequence *(ij)*, the criterion can be as follows:

$$t_{ij/\min} = \left\{(1-u_{ij})\left[w_{ij}t_{ij/h} + (1-w_{ij})t_{ij/m}\right] + u_{ij}t_{ij/v}\right\} \tag{5.8a}$$

where

u_{ij}, w_{ij} are elements of control functions **U** and **W** respectively, applied to select the separation rule between aircraft in sequence *(ij)*.

The control functions **U** and **W** are both in three-dimensional matrix form. The two dimensions-columns and rows-correspond to the number of different aircraft wake vortex categories. The third dimension represents the number of values which particular elements of both matrices can take. These are either 0 or 1.

Function **U** is at a higher hierarchical level than function **W**. This implies that an ATC controller using a convenient decision support tool, after deciding on the service discipline, can first select the value of **W** and then the value of **U**. Based on expression (5.8a), the values of **W** and **U** for the given aircraft sequence *(ij)* can be selected using the following criteria.

$$w_{ij} = \begin{bmatrix} 1, & \text{if } t_{ij/h} \le t_{ij/m} \\ 0, & \text{otherwise} \end{bmatrix} \quad \text{and} \quad u_{ij} = \begin{bmatrix} 1, & \text{if } t_{ij/v} < w_{ij}t_{ij/h} + (1-w_{ij})t_{ij/m} \\ 0, & \text{otherwise} \end{bmatrix} \tag{5.8b}$$

where all symbols are as in the previous expressions.

In addition, the values of **W** and **U** can be selected depending on the time of day and other local conditions.

5.5 Application of Proposed Methodology

5.5.1 *Background*

The methodology for assessing the effects of the selected advanced operational procedures, regulations, and technologies on increasing airport runway landing capacity is applied to the generic case of a single runway using the "what-if" scenario approach [16–18].

5.5.2 *The ATC Time-Based Separation Rules*

5.5.2.1 Input

The model of the "ultimate" capacity of a single runway using time-based instead of distance-based ATC separation rules between landing aircraft is applied using generic input. This relates to the size (i.e. geometry) of the "wake reference airspace", characteristics of the wake vortices of the landing aircraft fleet, behaviour of the wake vortices within and around the "wake reference airspace" influenced by external weather conditions such as crosswind and headwind, and ATC distance-based separation rules.

Size of the "Wake Reference Airspace"

The size of the "wake reference airspace" is determined by using the following input: the length of the common approach path between FAG and the runway landing threshold T is deemed to be similar to that at most airports, i.e. $= 6$ nm. Since aircraft use ILS, the distance from the threshold to the ultimate point of touchdown is assumed to be $\Delta = 0.16$ nm, i.e. 300 m. This gives the total distance between the FAG and runway touchdown of 6.16 nm. The nominal ILS GS angle is $\theta = 3°$ with maximum deviations of about $\pm 0.5°$. The angle between the axis and each side of the "wake reference corridor" in the horizontal plane is determined by the characteristics of the ILS LLZ (Localizer) and amounts to $\alpha = \pm\ 1.5°$. The distance between the ILS Outer Marker (OM) and the landing threshold T is $\beta = 4$ nm. Consequently, "wake reference profiles" are calculated depending on the distances and times from the landing threshold and are given in Table 5.3.

Characteristics of the Aircraft Fleet

The aircraft types are categorised into four categories according to Table 5.1. Their average characteristics, based on the specific values of particular parameters

Table 5.3 Size of the "wake free profile" depending on the distance and the time to the landing threshold

Distance to landing threshold (nm)/(s)[a]	Size of the profile	
	y (ft)	z (ft)
6/0	2,000	600
5/27	1,600	500
4/54	1,200	400
3/81	950	300
2/108	640	200
0/162	200	50

[a] Based on an average aircraft speed of 135 kts

Table 5.4 Characteristics of particular aircraft landing categories (averages values)

Aircraft category	Mass M (10^3 kg)	Wing span B (m)	Approach speed V (kts)[a]	Circulation Γ_0 (m/s^2)[b]	Wake reference time t^*(s)[b]	Glide Slope angle θ(°)
Small	20	24	120/90	138/184	16/12	3/4/5.5
Large	55	30	140/120	260/303	13/12	3/4/–
B757	117	38	170/140	359/436	16/13	3/4/-
Heavy	206	65	170/140	370/449	44/36	3/-/-

[a] The maximum and minimum approach speed, respectively, at the FAG and landing threshold T
[b] The values correspond to the maximum and minimum approach speed, respectively
Compiled from [20, 21]

including the calculated wake vortex parameters of each particular category, are given in Table 5.4.

In addition, the initially generated wake vortices are assumed to decay to the observed typical atmospheric background circulation of $\Gamma^* = 70$ m^2/s over the period $k = 8t^*$ [25, 26]. The proportion of particular aircraft categories in the aircraft fleet mix is varied parametrically.

External Conditions

External conditions are characterised by a constant crosswind of $V_{\mathrm{cw}} = 5$ m/s, which is above the conditions of "no wind" of $V_{\mathrm{cw}} \leq 3$ m/s. The influence of headwind $V_{\mathrm{hw}}(t)$ is not particularly considered since some preliminary calculations have shown that even very strong headwind cannot increase the vertical distance between the wake vortex of the leading and the flight path of the trailing aircraft in a shorter time than that obtained by the current ATC distance-based separation rules.

ATC Separation Rules

The minimum ATC distance-based separation rules in Table 5.1 are used as the basis for initially setting up time-based separation rules in combination with the average runway landing occupancy time of $t_{\mathrm{ai}} = 60$ s for all aircraft categories.

5.5.2.2 Results

The results of applying the model consist of the following components:

- The strength (i.e. circulation) of wake vortices to which the trailing aircraft are exposed in particular landing sequences if ATC VFR and IFR from Table 5.1 are applied;
- The matrix of standardised time-based separation rules for particular categories of aircraft landing sequences; and
- The runway landing capacity calculated for current ATC distance-based VFR and IFR separation rules, and the wake vortex behaviour influenced by external weather (wind) conditions.

The strength (i.e. circulation) of the wake vortices to which the trailing aircraft in particular landing sequences are potentially exposed to when the minimum ATC IFR and VFR are applied is given in Table 5.5.

As can be seen and as intuitively expected, the potential wake vortex strength is higher under VFR than under IFR. In addition, in both cases and for most sequences, this circulation is significantly higher than the typical atmospheric circulation of 70 m^2/s. Furthermore, it should be borne in mind that the different types of trailing aircraft in the particular sequences are sensitive to the different strengths of the wake vortices in varying degrees. Last but not the least, trailing aircraft are not actually exposed to such circulation because the wakes of the leading aircraft sink below their flight paths thanks to their self-induced descent speed at the same time as they decay. This again illustrates the fact that the landing aircraft could also be put closer to each other under IMC as under VMC without significant risk of wake vortex hazard, providing, of course, that the corresponding technology for "see and be seen" is also available under IMC. In such a case, the separation rules under IMC and VMC would be unified as time-based separation rules. The basis for setting up these rules would be existing ATC VFR (Table 5.1) and typical aircraft approach speeds (Table 5.4). At least two positive effects could be achieved. Firstly, IFR capacity would generally increase and eventually equal the current VFR capacity. Secondly, the capacity would become much less sensitive to weather conditions, which would enable relatively stable runway operations and more predictable aircraft/flight delays. Table 5.6 gives an example of such standardised time-based separation rules.

The values in Table 5.6 are rounded-up for convenience in practical use. As can be seen, in some landing sequences, the runway landing occupancy times can be used as the minimum separation rules. In addition, as in the case of distance-based rules, these rules are applied depending on the landing sequence at the runway threshold ($v_i \leq v_j$) and at the FAG ($v_i > v_j$).

Using the above-mentioned inputs in Tables 5.4 and 5.6, the runway landing capacity is calculated for different cases.

Figure 5.5 shows the dependence of this capacity on the proportion of heavy aircraft in the fleet, the above-mentioned separation rules, and the wake vortex

Table 5.5 The potential circulation Γ(*t*), which the trailing aircraft faces under ATC VFR and IFR while flying at given approach speeds

i/j	Small	Large	B757	Heavy
ATC VFR				
Small	134	134	134	134
Large	207	231	231	231
B757	244	275	305	313
Heavy	317	333	379	379
ATC IFR				
Small	17	62	69	69
Large	0	87	101	101
B757	0	70	79	79
Heavy	181	234	261	197

Table 5.6 The standardised ATC time-based minimum VFR/IFR separation rules $\tau_{ij/min}$ for landing aircraft (min)

i/j	Small	Large	B757	Heavy
Small	1.0	1.0	1.0	1.0
Large	1.5	1.0	1.0	1.0
B757	2.0	1.3	1.5	1.2
Heavy	2.5	1.5	1.5	1.2

characteristics and behaviour. The proportion of small and B757 aircraft is constant at 5%. As can be seen in all cases, capacity decreases as the proportion of heavy aircraft in the mix increases up to about 20%, and then begins to increase again. In the former case, the impact of strong wakes behind heavy aircraft prevails, while in the latter case, the contribution of higher approach speeds of heavy aircraft prevails. In addition, the capacity for nominal ATC VFR is higher than the capacity for nominal ATC IFR by about 30%, as also shown in Fig. 5.2. If time-based separation rules were applied under conditions of crosswind of 5 m/s, the capacity would be somewhere between the current VFR and IFR capacity. This indicates that the capacity gains would be comparable to IFR capacity if the influence of crosswind on wake vortex behaviour is taken into account. When time-based separation rules are applied under conditions of respecting the wake vortex descent time, the corresponding capacity will be lower than the current IFR capacity. This implies that the current IFR seem to be based only partially on the descent time of the wake vortices below the flight path of the trailing aircraft and not on the time they take to completely move out of the "wake reference airspace".

Dynamically selected time-based separation rules for particular landing sequences combining current ATC VFR and ambient factors influencing wake vortex behaviour seem to give the highest capacity. However, in the given example, this capacity is only slightly higher than the capacity obtained under the current ATC VFR. This again suggests that the current distance-based ATC VFR could be the basis for setting up corresponding time-based separation rules applicable to both VMC and IMC.

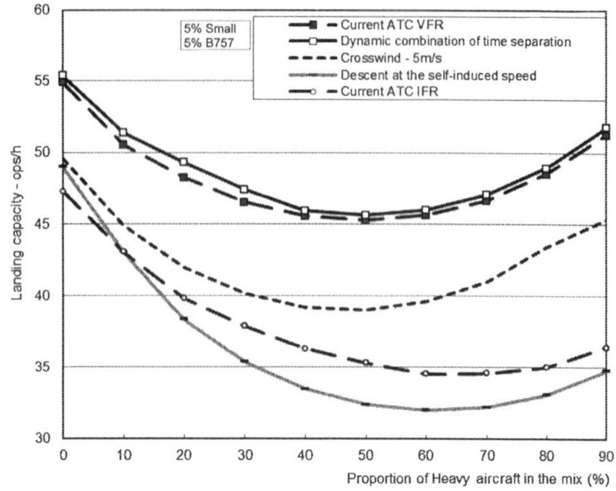

Fig. 5.5 Landing capacity of a single runway depending on the aircraft fleet mix, ATC separation rules, and the wake vortex characteristics and behaviour (Compiled from [16])

5.5.3 Prioritising Aircraft Landings

5.5.3.1 Input

The input on ATC separation rules from Table 5.1, the average aircraft approach/ speeds from Table 5.4, and the length of the final approach path are used for calculating the "ultimate" landing capacity of a single runway when FCFS and PR service rules are applied [18].

5.5.3.2 Results

The results are shown in Fig. 5.6. As can be seen, the "ultimate" capacity from both service rules decreases as the proportion of heavy aircraft in the mix increases. Specifically, for the mix with 10% of small aircraft, the service rate is higher for the PR than for the FCFS rule by about 7–21%, depending on the proportion of heavy aircraft in the arrival mix. If the mix consists of only large and heavy aircraft, this difference ranges from 0 to about 15%, reaching the highest value if the mix contains 50% of heavy aircraft. In both cases, this difference is greatest if the proportion of heavy aircraft in the mix varies between 40 and 60%. At most large airports, this proportion amounts to about 20–25%. This implies a potential difference in "ultimate" landing capacity of about 10–15% if the PR rule is applied instead of the FCFS priority rule. Consequently, it appears that prioritising of aircraft landings would be more effective in cases of a greater heterogeneity of the aircraft fleet mix than otherwise.

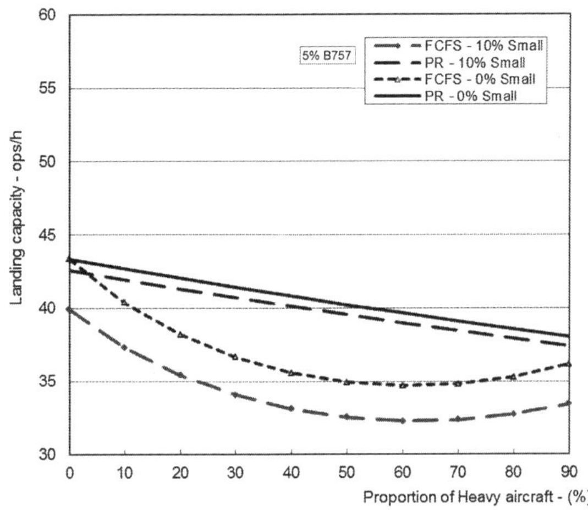

Fig. 5.6 Landing capacity of a single landing runway for the FCFS and the PR rule (Compiled from [18])

5.5.4 The ATC Vertical Distance-Based Separation Rules

5.5.4.1 Input

The ATC IFR separation rules in Table 5.2 are used to calculate the runway landing capacity as the benchmarking case. In addition, the standardised time-based separation rules in Table 5.6 are used for comparison [18], and the minimum ATC vertical distance-based separation rules of 1,000 ft are assumed to be applied to all sequences of landing aircraft.

The characteristics of aircraft types including the assumed GS angles are given in Table 5.4. As can be seen, according to these assignment scenarios, small, large, and B757 aircraft are assumed to approach and land at more than one GS angle while heavy aircraft will continue to exclusively use a GS angle of 3°. The fleet mix is varied while maintaining the proportion of small and B757 aircraft constant in all cases (5%).

Based on GS in Table 5.4 and respecting the two cases when the mixture of ATC horizontal and vertical distance-based separation rules and exclusively ATC vertical distance-based separation rules are applied, the combinations of GS angles for particular landing sequences are given in Tables 5.7 and 5.8.

Specifically, in the case when a mixture of ATC horizontal and vertical distance-based separation rules is applied, i.e. when approach and landing is carried out along "Flight Corridors", the length of common approach path with a GS angle of 3° for all aircraft is adopted to be: $\gamma = 6.16$ nm [16]. The same length of approach path is used when the capacity is calculated for the case of using exclusively ATC vertical separation rules.

Table 5.7 GS angles when a mixture of the ATC minimum mixed horizontal/vertical distance-based separation rules are applied (°)

i/j	Small	Large	B757	Heavy
Small	3/3	3/3	3/3	3/3
Large	3/5.5	3/3	3/3	3/3
B757	3/5.5	3/4	3/3	3/3
Heavy	3/5.5	3/4	3/4	3/3

Table 5.8 GS angles when the ATC minimum vertical distance-based separation rules are applied (°)

i/j	Small	Large	B757	Heavy
Small	3/5.5	3/4	3/4	3/3
Large	3/5.5	3/4	3/4	3/3
B757	3/5.5	3/4	3/4	3/3
Heavy	3/5.5	3/4	3/4	3/3

5.5.4.2 Results

The results from applying the model using the above-mentioned inputs are given in Figs. 5.7 and 5.8.

Figure 5.7 shows the dependency of the runway landing capacity on the proportion of heavy aircraft in the fleet mix and type of ATC minimum separation rules when FCFS service discipline is applied. As can be seen and as already shown in Figs. 5.6 and 5.7, when the current ATC IFR separation rules and time-based separation rules are applied, landing capacity decreases as the proportion of heavy aircraft in the fleet increases. When ATC mixed horizontal and vertical distance-based separation rules are applied, the capacity increases with the proportion of heavy aircraft in the mix. The main reason is that vertical separation applied to FS (Fast–Slow) sequences at the entry gate reduces horizontal distances between the aircraft, thus shortening the corresponding inter-arrival times at the landing threshold and consequently contributing to increased capacity. When only ATC minimum vertical distance-based separation rules are applied, capacity again decreases as the proportion of heavy aircraft in the mix increases. This occurs because these aircraft use smaller GS angles that do not allow shortening of horizontal distances in particular aircraft sequences, which despite higher approach speeds increase the corresponding inter-arrival times at the landing threshold and consequently decrease capacity. In addition, the capacity is greater by between 2 and 33% in the case when vertical distance-based separation rules are exclusively applied than in the case when a mixture of ATC horizontal and vertical distance-based separation rules is applied; however, the difference decreases as the share of heavy aircraft in the mix increases. At the same time, purely ATC minimum vertical separation rules enable achievement of the greatest landing capacity for the widest range of proportion of heavy aircraft in the mix, followed by ATC time-based, ATC mixed, and finally ATC IFT horizontal-distance-based separation rules.

Figure 5.8 shows the dependence of runway landing capacity on the proportion of heavy aircraft in the mix, ATC minimum mixed horizontal and vertical

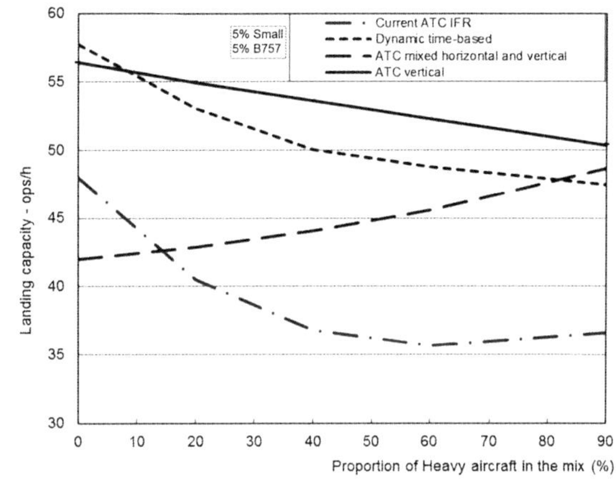

Fig. 5.7 Landing capacity of a single runway depending on the proportion of heavy aircraft in the mix and different types of ATC separation rules—FCFS service discipline

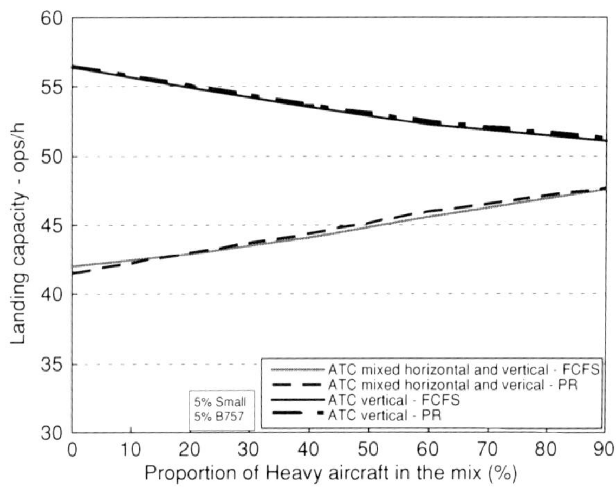

Fig. 5.8 Landing capacity of a single runway depending on the proportion of heavy aircraft in the mix, and different types of ATC separation rules—FCFS and PR service discipline

distance-based and exclusively vertical distance-based separation rules, and service disciplines FCFS and PR. As can be seen, for both service disciplines, the distinction between capacity when ATC mixed and exclusively vertical distance-based separation rules are applied remains as the proportion of heavy aircraft in the mix increases. However, in both cases, switching service discipline from FCFS to PR does not significantly increase the runway landing capacity.

5.6 Concluding Remarks

This chapter describes the prospective effects of some advanced operational procedures and regulations supported by existing and/or advanced technologies for increasing the "ultimate" landing capacity of the airport runway. These include: (1) ATC time-based separation rules; (2) prioritising; and (3) ATC mixed horizontal and vertical, and exclusively vertical distance-based separation rules applied between aircraft landing on a single runway. The latter approach is possible due to the assumed capability of aircraft to safely approach and land on a given runway at different flexible ILS/MLS GS (Glide Slope) angles.

A methodology consisting of dedicated models of "ultimate" runway landing capacity under the above-mentioned conditions has been developed and applied to the generic case of a single runway according to the "what-if" scenario approach.

The model of landing capacity based on time-based separation rules has been applied to a single runway with specified geometry of the "wake reference airspace". The runway is assumed to serve four FAA/ICAO aircraft categories, which are characterised by the wake vortex parameters (approach speed, wing span, and weight), and the runway landing occupancy time under given atmospheric conditions (crosswind). The results have shown that dynamically selected time-based separation rules based on the current ATC VFR and influence of the crosswind on wake vortices give the highest runway landing capacity. Time-based separation rules based on the wake vortex's self-induced descent speed produce a landing capacity slightly lower than that achieved under current ATC IFR. In these cases, landing capacity generally decreases as the heterogeneity of the aircraft fleet mix increases, and is generally lower as the proportion of heavy aircraft in the fleet increases.

The model of landing capacity under prioritising aircraft landings has also been applied to a single landing runway. The results indicate that the (PR) priority serving rule could substantially increase the runway "ultimate" landing capacity compared to the current FCFS (First-Come–First-Served) priority rule. The results also confirm that prioritising aircraft landings can only be beneficial if applied to a heterogeneous aircraft fleet. In addition, if consistently applied under the specified conditions, prioritising could act similarly to a congestion charging measure by deterring access of smaller aircraft and stimulating an increase in the average aircraft size. An alternative that could be considered would be to introduce the priority criteria of an essentially administrative character, such as those at NY LaGuardia airport where prioritising is based purely on economic criteria, but this seems to be inefficient primarily due to a rather homogeneous aircraft fleet.

Finally, the model of landing capacity based on application of ATC mixed horizontal and vertical, and exclusively vertical separation rules, has been applied to a single runway using the other inputs from the other two models. The results indicate that the runway landing capacity could be substantially increased by applying ATC minimum vertical distance-based separation rules when the ATC FCFS service discipline is applied. This capacity is the greatest as compared to the capacity obtained by applying ATC time-based, ATC mixed horizontal and vertical distance-based, and exclusively ATC horizontal distance-based separation rules. In addition, this capacity decreases at a decreasing rate as the proportion of heavy aircraft in the fleet mix is increased, which also applies to its differences as compared to the capacities, which would be achieved by application of other combinations of ATC separation rules. In addition, when ATC mixed horizontal and vertical and exclusively vertical distance-based separation rules are applied, the capacity does not significantly change with service discipline, i.e. FCFS or PR.

In general, the above-mentioned results indicate that advanced operational procedures and rules supported by improved existing and advanced new technologies could increase the runway "ultimate" landing capacity by about 20–30% without the need for building new runways requiring additional land take (use) and consequent increase in airport size. Thus they could contribute to greening, i.e. more sustainable development of particular spatially constrained airports at least in the short- to medium-term period.

References

1. Andreatta G, Romanin-Jacur G (1987) Aircraft flow management under congestion. Transp Sci 21(4):249–253
2. Blumstein A (1959) The landing capacity of a runway. Oper Res 7(1):752–763
3. Choroba P (2003) A vision of wake vortex research for next 20 years. Working paper, EUROCONTROL, Bretigny sur de Orge, Paris, France
4. Czerny AI, Forsyth P, Gillen D, Niemeier HM (2008) Airport slots: international experiences and options for reform. Ashgate Publishing Limited, Aldershot
5. Donahue GL, Shorttle J, Jeddi B (2008) Looking for the capacity in NGATS. On Transportation Research Board (TRB) annual meeting, Washington DC, USA, 14 Jan 2008
6. EC (2002) Study on competition between airports and the application of state aid rules, volume II. Air Transport Group, Canfield University, European Commission DG VII-F, Brussels, Belgium
7. EC (2003) Analysis of the European transport industry. European Commission–DG TREN, Final report, Brussels, Belgium
8. EC (2005) Optimal procedures and techniques for the improvement of approach and landing–OPTIMA, Sixth framework program, FP6-2002-Aero 1502880, Deliverable 1.2, European Commission, Brussels, Belgium
9. EC (2007) Gaps identification airport, TMA, En route and TBS–reduced separation minima-RESET. Deliverable D4.2, Sixth framework program, European Commission, Brussels, Belgium, p 86
10. FAA (2004) Summary results from long-term wake turbulence measurement at San Francisco international airport. U.S. Department of transportation, Federal Aviation Administration, DOT-VNTSC-FA27-PM-04-13, Washington, DC, USA
11. FAA (2010). http://www.faa.gov/about/initiatives/nextgen/
12. Gilbo EP (1993) Airport capacity: representation estimation, optimization. IEEE Trans Control Syst Technol 1(3):144–153
13. Harris RM (1972) Models for runway capacity analysis. The MITRE Corporation technical report, MTR-4102, Rev. 2, Langley, Virginia, USA
14. Janic M, Tosic V (1982) Terminal airspace capacity model. Transp Res 16A(4):253–260
15. Janic M (2001) Air transport system analysis and modeling: capacity, quality of service and economics. Transport series books, vol 16. Taylor and Francis, UK
16. Janic M (2008) Towards time-based separation rules for landing aircraft. Transportation Research Record (TRB), No. 2052 Aviation, pp 79–89
17. Janic M (2008a) Modeling the capacity of closely-spaced parallel runways using innovative approach procedures. Transp Res C 16(6):704–730
18. Janic M (2009) A concept for prioritizing the aircraft operations at congested airports. Transportation Research Record (TRB), No. 2106 Aviation, pp 100–108

19. Newell GF (1979) Airport capacity and delays. Transp Sci 13(3):201–241
20. NASA (1999) Benefit estimates of terminal area productivity program technologies. NASA/CR–1999 208989. National Aeronautics and Space Administration, VA
21. NASA (2001) Enhancement airport capacity through safe dynamic reduction of aircraft separation: NASA's aircraft vortex spacing system (AVOSS). National Aeronautical and Space Administration, Hampton
22. Odoni AR, Bowman J (1997) Existing and required modelling capabilities for evaluating ATM systems and concepts. Final report no. NAG2-997, International Centre for Air Transportation. Massachusetts Institute of Technology, Massachusetts, Boston, USA
23. RITA (2008) Airport statistics 2007. The Research and Technology Innovative Administration, US Department of Transportation, Washington DC. http://www.transtat.bts.gov
24. Robinson III JE, Davis TJ, Isaacson DR (1997) Fuzzy reasoning based sequencing of arrival aircraft in the terminal area. The American Institute of Aeronautics and Astronautics (AIAA) Guidance, Navigation, and Control Conference, New Orleans, USA, p 11
25. Sarpkaya T (2000) New model for vortex decay in the atmosphere. J Aircr 37(1):53–61
26. Shortle JF, Jeddi BG (2007) Probabilistic analysis of wake vortex hazards for landing aircraft using multilateration data. 86th TRB Conference, Transportation Research Board, Washington DC, USA
27. Swedish WJ (1981) Upgraded airfield capacity model. The MITRE Corporation Technical Report, MTR-81W16, Vols 1 and 2. Langley, Virginia, USA
28. Terrab M, Odoni AR (1993) Strategic flow management in air traffic control. Oper Res 41(1):138–152

Chapter 6
Greening the Airport Airside Area II: Liquid Hydrogen as an Alternative Fuel

6.1 Introduction

Global air traffic has increased from 0.5 trillion RPK (Revenue Passenger Kilometres) in 1971 to about 4.25 trillion RPKs in 2006. Some long-term forecasts by international air transport organisations (IATA, ICAO, ACI), and in particular by the two main manufacturers of commercial aircraft Boeing and Airbus, predict rather stable RPK growth at an average annual rate of 5% over the next 20 years. This will increase the total volumes of the world's traffic to about 10.545 trillion RPKs [2] or 11.4 trillion RPKs [5] by the year 2025/2026. At the same time, the number of passengers is predicted to rise at an annual rate of 4.5%, which will result in their total number of about 6.8 billion in 2025/2026 [5]. In addition, air cargo traffic is forecasted to increase at an average annual rate of 6.1% over the same period, from about 200 billion RTKs in 2006 to about 650 billion RTKs in 2025/2026 [5]. Such growth will require an increasing number of aircraft, from the current 18,230 (of which 16,250 are passenger aircraft) in 2006 to about 36,420 (of which 32,440 will be passenger aircraft) in 2025/2026 [5]. Since all these aircraft continue to be powered by conventional jet fuel, a derivative of crude oil, the cumulative fuel consumption and related emissions of the main greenhouse gases such as CO (Carbon-Oxide), CO_2 (Carbon-Dioxide), SO_4 (Sulphur-Oxides), NO_x (Nitrogen-Oxides), H_2O (water vapour), and particles will continue to increase, thus contributing to depleting crude oil[1] reserves on the one hand, and global warming and climate change on the other [2, 5, 16, 19]. Some estimates indicate that air transport emitted about 513 $MtCO_2$ in 1992, is expected to increase to about 1,468 $MtCO_2$ in 2050. Of this total, airports account for and will continue to account for only about 5%, i.e. 30 million tons per year. The above-mentioned total amount of

[1] Global reserves of crude oil are estimated at between 854 and 1,255 Gb (Giga barrels), which according to average consumption of 30.3 Gb/day provides for consumption over the next 30–40 years [16].

M. Janić, *Greening Airports*, Green Energy and Technology
DOI: 10.1007/978-0-85729-658-0_6, © Springer-Verlag London Limited 2011

CO_2 emitted by the air transport system will likely continue to account for between 3 and 5.5% of the total CO_2 emissions by man [1, 20].

Such prospective increasing of greenhouse gases at both the global-system and local—airport scale has raised the question of short, medium- and long-term viable mitigating alternatives, and related tactics and strategies for creating a potential "carbon–neutral"[2] or "purely green" air transport system and its components–airports [14, 19, 22, 23].

Currently, the following alternatives, tactics and strategies for airports have been discussed [1]:

- Preventing access to highly polluting aircraft;
- Reducing aircraft fuel consumption during the LTO (Landing and Take Off) cycle and preventing use of APUs (Auxiliary Power Units) at the apron/gate stands;
- Reducing the overall number of vehicles accessing and operating at a given airport;
- Encouraging the use of low and/or zero emission vehicles within the airport area;
- Reducing energy consumption in all buildings;
- Implementing different schemes of charging externalities such as emission trading or taxation;
- Developing true multimodal transport nodes; and
- Stimulating use of alternative fuels.

The first six alternatives are self-explanatory and the seventh alternative is elaborated in Chap. 4. The final alternative is the subject of this chapter, but dealt with in a wider scope. This implies considering the potential contribution of alternative fuels to long-term "greening", i.e. more sustainable development, in terms of fuel consumption and emissions of greenhouse gases both at the global–air transport system and the local–airport scale. Specifically, the influence of gradually replacing conventional Jet A fuel—kerosene with an alternative fuel—LH_2 (Liquid Hydrogen)—on the cumulative emission of greenhouse gases, i.e. air pollution, by both the air transport system as a whole and a given airport as its components, is described.

6.2 Fuels, Aircraft, and Airport Fuel-Supply/Storage Systems

6.2.1 Fuels

6.2.1.1 Conventional Jet A Fuel–Kerosene

Conventional Jet A fuel–kerosene, which powers most of today's commercial jet aircraft, is a derivative of crude oil [33]. The most important characteristics of this jet fuel are its specific energy (MJ/kg) and volumetric energy (MJ/dm^3)

---

[2] The concept of a "carbon–neutral" air transport system implies its further growth without cumulative emissions of greenhouse gases increasing, stagnating and/or increasing at a decreasing rate [21, 26].

content. In general, a given quantity of fuel with a higher specific energy enables an aircraft to carry more passengers and cargo over a given distance. In addition, such an aircraft can cover longer distances while carrying a given number of passengers and cargo. In addition, fuels with a higher volumetric content enable aircraft to fly longer distances, and vice versa. The main greenhouse gases emitted by burning conventional Jet A fuel are mentioned above. The emission rates of the two most important greenhouse gases, CO_2 and H_2O, are constant at 3.18 g/g of Jet A fuel, and 1.26 g/g of Jet A fuel, respectively. The emission rates of the 3rd most important greenhouse gas, NO_x, mainly depend on the fuel burning temperature. For example, this rate is approximately 0.015 g/g of Jet A fuel during the cruising phase of flight (see also Chaps. 2 and 3) [19, 21].

6.2.1.2 Liquid Hydrogen

Sources and Logistics

Manufacturing: What is LH_2 as an aviation fuel? It is the H_2 gas in liquid state. In order to obtain this liquid, gas is compressed and then cooled to under −217°C, after which point it converts into liquid. Some methods and processes for manufacturing LH_2 are already commercially available, but used only in a small niche market as a chemical substance and not as an energy source. For general commercial use, LH_2 can be produced from chemically reformed natural gas, fossil fuel, and/or biomass feedstock by using conventional chemical processes. In addition, it can also be produced by dissociating water using electricity, heat, sunlight, and/or specialised micro-organisms. In these cases, its production will mainly be driven by economic reasons including full logistics costs [11, 18].

Transport and storage: LH_2 can be transported and stored after conversion into a highly concentrated form by increasing the pressure and/or by lowering the temperature. In general, over shorter distances, it is transported as a compressed gas and also through the pipeline system. Over longer distances, it is transported as a liquid by dedicated vehicles operating on all transport modes. It is stored using high-pressure cylinder cryogenic tanks and containers [17].

Economy: The "economy" of LH_2 as a prospective fuel for commercial aircraft implies the amount of energy consumed for its production, packaging, transport, and storage, all depending on whether it is in the liquid or gaseous state. For example, the energy input could be between 2.12 and 1.65 times higher than the energy content of the delivered liquid and gas hydrogen, respectively (i.e. loss factor). In both cases the loss factor is considerably higher than that of conventional jet fuels, where it amounts to around 1.12.

Costs/price: The most important issue in supplying LH_2 as an energy commodity is its competitive price. This depends greatly on the raw material used and the related manufacturing processes on the one hand, and some market mechanisms such as for example taxes on CO_2 emissions on the other. In general, the

Table 6.1 Characteristics of conventional Jet A fuel and LH_2

	Type of fuel	
	Jet A fuel-kerosene	LH_2
Burning characteristics		
Specific energy (mJ/kg)	43.2	120
Specific density 15°C(kg/dm^3)	0.790–0.808	0.071
Energy density (mJ/l)	34.9	8.4
Boiling point (°C)	167–266	–253
Emissions[a]		
CO (g)	0.50	–
CO_2 (kg)	0.75	–
H_2O (kg)	0.30	0.75
NO_x (g)	0.41	0.02–0.102
UHC (g)	0.20	–

[a] Based on 10 mJ of energy content obtained from 0.5 l LH_2 and 0.3 l of Jet A fuel. Compiled from [11]

price of hydrogen should be comparable to that of conventional jet fuel at that time. In such contexts, prices for conventional jet fuel are expected to rise and that of hydrogen to decrease in the long term, which makes the expectation of comparable prices for both fuels more realistic. Some estimates indicate that in 2035, the production costs of hydrogen will range between 0.8 and 3.5 \$US/kg H_2 [18].

Operational Characteristics

The main operational characteristics of hydrogen as a jet fuel are its specific energy (120 mJ/kg), specific density (0.071 kg/dm^3 at 15°C), energy density (8.4 mJ/l), and boiling point (–253°C) [11]. These characteristics point out the requirements for specific aircraft design and the overall logistics of hydrogen as jet aviation fuel.

Environmental Characteristics

Liquid Hydrogen as a jet fuel has the following advantages in terms of emissions of particular greenhouse gases compared to conventional jet fuel: 0 vs. 0.50 g of CO; 0 vs. 0.75 kg of CO_2; 0.78 vs. 0.30 kg of H_2O; 0.02–0.102 vs. 0.41 g of NO_x; and 0 vs. 0.20 g of UHC (this is based on 10 mJ of energy content obtained from 0.5 l of LH_2 and 0.3 l of Jet A fuel). Thus, the burning of LH_2 does not produce CO_2 and SO_x. The only matter of concern is increased H_2O emissions. Table 6.1 Summarizes the above-mentioned characteristics of both Jet A and LH_2 as aviation fuels.

As can be seen, in addition to the above-mentioned superiority in energy performance, hydrogen directly emits drastically lower amounts of all greenhouse gases as compared with emissions from conventional jet fuel, with the exception of H_2O (about 2.6 times more).

Safety

Liquid Hydrogen is considered a safe fuel. Nevertheless, its main potential disadvantages are its explosive rate of 13–79% concentration in the air and its very low ignition energy (as low as about 0.02 mJ). Hydrogen also mixes with air faster than jet fuel vapour, and disperses rapidly through the air, in contrast to jet fuel which pools on the ground. It burns with a nearly invisible, colourless, and odourless flame, which is also an important safety concern [18].

6.2.1.3 Cryogenic Commercial Aircraft

State of the Art Development

The first ideas and related experiments with LH_2 as a prospective fuel for commercial air transport emerged over 70 years ago. However, the first commercial aircraft powered by LH_2 was based on the USSR's Tupolev TU-154 and built in 1988 as an experimental aircraft [30]. In order to use LH_2, some modifications had to be made to the airframe, fuel supply and storage systems, and the aircraft engines (NK88). In addition, systems ensuring fire/explosion safety and data recording were also installed. From 15 April 1988, the aircraft successfully made several flights between the airport of Moscow (Russia) and other European airports including Bratislava (Czechoslovakia), Nice (France), and Berlin and Hanover (Germany). However, as it had started, the program suddenly stopped due to changed priorities relating to Liquid Natural Gas (LNG) as a prospective aviation fuel as well as due to the political and economic crisis in the USSR which began in 1989.

Currently, research on using LH_2 for commercial air transportation is underway in Europe, the US, and the Russian Federation. In Europe, various projects including system analysis of the feasibility of LH_2 powered aircraft, their design, and scenarios for their eventual implementation have been completed [15]. They have provided foresight into the prospective technical/technological, operational, economic, environmental, and safety characteristics of cryogenic aircraft, which are expected to be fully developed by around 2020 and consequently enter commercial service by around 2040 [15].

Design and Operational Characteristics

Regarding the characteristics of LH_2 compared to conventional jet fuel (2.8 times higher specific energy and about 11 times less specific density), cryogenic aircraft will require about 4.3 times more fuel volume for an energy output equivalent to conventional aircraft. Therefore, their main design characteristic will be the relatively large volume of their well-insulated cylindrical fuel tanks. They can assume various positions within the aircraft configuration: above the payload (passengers and freight), above and aft of the payload, and fore and aft of the payload section. Their wings, with no fuel storage space, could be smaller. This will result in increased aerodynamic resistance and aircraft empty weight compared to their conventional counterparts. However, the much lower weight of LH_2 is expected to compensate for such increase in the aircraft empty weight and consequently contribute to reducing the maximum take-off weight of cryogenic aircraft [10, 14].

Cryogenic jet engines will retain the basic structure of conventional jet engines, albeit with some necessary modifications, such as fuel pumps, fuel control units, and combustion chambers. Experiments so far have shown that such cryogenic engines will have about 64% lower Specific Fuel Consumption (SFC) than conventional jet engines (0.0976 vs. 0.2710 (kg/h)/kg for cruising and 0.0512 vs. 0.1420 (kg/h)/kg for the take-off phase of flight). In addition, these engines are expected to be 1–5% more efficient in generating thrust from the given energy content. For Supersonic Transport Aircraft (STA), the specific consumption of LH_2 and Jet A fuel during cruising is expected to amount to about 0.260 (kg/h)/kg and 0.680 (kg/h)/kg, respectively (the ratio Jet A/LH_2 is 2.61). Last but not the least, hydrogen engines for either aircraft type are expected to operate with a slightly lower turbine entry temperature, which in turn extends their useful life and reduces maintenance costs [10, 14].

Economic Characteristics

During 2003–2007, the share of fuel costs in the total airline operating costs amounted to around 30% [15]. Considering the unit price of LH_2, the latest price of conventional jet fuel, and the lower Specific Fuel Consumption of cryogenic engines of about 64% (i.e. 1 kg Jet A is equivalent to 0.36 kg LH_2), estimations show that the share of fuel costs in total operating costs of cryogenic aircraft could vary between 45% (1$US/kg LH_2) and 78% (1.73$US/kg LH_2) (the prices of other inputs are assumed constant). Equalising the prices of both fuels to 1$US/kg, the shares of corresponding costs would amount to about 60 and 35%, respectively, mainly due to the lower Specific Fuel Consumption of cryogenic aircraft. This scenario appears quite realistic since the prices of conventional jet fuel are expected to continue to rise, while the price of LH_2 is assumed to decrease as the efficiency of its manufacture, storage, and distribution improves.

Environmental Characteristics

Cryogenic aircraft, powered by LH_2, do not emit CO_2. However, water vapour (H_2O) emitted in quantities of about 2.6 times higher than conventional aircraft at

Table 6.2 The relative characteristics of typical long-range conventional and cryogenic aircraft

Attribute	Conventional aircraft (Jet A)	Cryogenic aircraft (LH_2)
Fuel energy content	1	0.36
Volume of fuel	1	11
Volume of fuel tanks	1	4.3
MTOW	1	0.85–1.05
Aerodynamic resistance	1	1.1
Pollutants CO, CO_2, SO_x, HC	1	0
H_2O	1	2.6
NO_x	1	0.05–0.25

Compiled from [15]

and above the cruising altitudes of about 31,000 ft (ft—feet; 1 ft = 0.305 m) will be the main greenhouse gas. That said, the impact of H_2O compared to greenhouse gases emitted by conventional aircraft appears to be much lower. Cruising at lower altitudes may be one option to counter this effect, although this would impact other performances. In addition, cryogenic aircraft emit about only 5–25% of NO_x compared to their conventional counterparts, a gain which is expected to be achieved through adequate design of the combustion chambers of cryogenic engines [15]. Table 6.2 Summarises the main differences between conventional and cryogenic aircraft.

Safety

Cryogenic aircraft should be as safe as conventional aircraft. In the case of an aircraft accident, LH_2 burns much faster (15–22 s) and with low heat radiation, thus mitigating the impact of fuselage collapse, which contrasts to the impact of burning conventional jet fuel. In addition, burning LH_2 covers a much smaller surface area [15]. The overall safety figure also includes appropriate design of the airport fuel supply system. It seems likely that LH_2 production will take place in the airport fuel storage area, reserves will be stored in large storage tanks, and the fuel will be delivered to the aircraft at the airport parking stands through a dedicated pipeline system.

6.2.2 Airport Fuel–Supply Systems

The main characteristics of the fuel-supply system(s) at a given airport generally depend on the volume of demand for particular types of aviation fuels. This implies the number of atms (air transport movements) to be carried out during a given period of time, and the structure of aircraft fleet in terms of aircraft size, type of flight (short, medium, long haul), and type of fuel used. In the case of a given airport, after starting operations with cryogenic aircraft, two fuel-supply systems will have to be installed and operated independently: one for conventional kerosene and another for LH_2.

6.2.2.1 Conventional Fuel

The system for supplying aircraft with conventional jet. A fuel-kerosene has become relatively large and complex task at large airports. The specific requirements, which appear to be increasingly important, are simultaneous fuel supply for many aircraft at their apron/gate parking positions and the limited time for supply within the aircraft turnaround time. Currently, at most airports, fuel is stored in the fuel tank area or the fuel farm located sufficiently far from the closest object(s) in the airport area. The storage capacity of the tanks usually allows continuous fuel supply for 1–3 days depending on the airport and the reliability of the fuel delivery from the fuel manufacturing plant, i.e. oil refinery (usually by rail and rarely by trucks). From the fuel farm, fuel is delivered to the aircraft at their apron/gate parking stands through a network of underground pipelines and/or by trucks. At large airports, underground pipelines are mostly used as a safer alternative and as solution for mitigating congestion at the apron/gate complex during peak periods. Fuel trucks are mainly used at smaller regional airports [24].

6.2.2.2 Liquid Hydrogen

The introduction of cryogenic aircraft can be commercially feasible only if a sufficient number of airports are equipped with corresponding fuel supply systems. The dynamism of installing such fuel supply facilities at particular airports will mainly depend on the dynamism of introducing cryogenic aircraft fleets. If such fleets will mainly consist of large-heavy aircraft, large airports hosting such aircraft will be the first to install such fuel supply systems.

The preferred alternative for supplying LH_2 is to produce it on-site at the airport mainly by electrolysis. This seems to be the optimal alternative regarding the distribution of LH_2 to the aircraft. In any case, the liquefaction plant should be located well away from runways' centrelines. Both the production and storage capacity of LH_2 will generally depend on the volume of peak daily demand. Since the liquefaction units have a limited production capacity, at airports with large daily demand, the production plant will have to be constructed modularly, i.e. consist of several modules. In addition, the storage tanks will have to have sufficient volume to handle that amount of production and provide some fuel reserve in case of potential failure of the production module(s). The storage tanks will be cylindrical and positioned vertically at a certain distance from each other for safety reasons. In addition, they will have to be far away from the liquefaction plant, which is a compromise between the required level of safety and losses in transferring LH_2. The area of land occupied by the liquefaction plant will mainly depend on the size of the plant, which in turn depends on the volume of demand per given unit of time (day).

Furthermore, in the case of coexistence with conventional Jet A fuel-supply systems, any LH_2 system should be located in a completely different area, but again sufficiently far away from the closest object(s) in the airport area. This implies that additional land will be needed for installing systems, which will be connected to the apron/gate aircraft parking stands via a dedicated network of pipelines. In order to maintain LH_2, the temperature inside the system has to be maintained at or below –253°C (see Table6.1). This means that the distribution pipes will have to be well-insulated and as short as possible, which can be achieved by using special steel. Each pipeline should be triplicated, which again for economic reasons means that their length should be as short as possible. In this context, the first primary pipeline will be used for distributing LH_2 to the aircraft. The second pipeline will be used for collecting gaseous hydrogen created by any cause, and its return to the liquefaction plant, mainly for safety and economic reasons. The final pipeline shall be used as a redundancy line for the distribution of LH_2 and the conversion of gaseous hydrogen. These pipes will end at the aircraft apron/gate/stands with the hydrant pits. In order to enable flexible use of particular parking gates/stands by both conventional and cryogenic aircraft, hydrants will be provided for both conventional and cryogenic fuel service. Alternatively, as in the case of supplying conventional Jet A fuel-kerosene, fuelling-trucks, and/or mobile tank-trucks can be used for LH_2. Both systems will allow gaseous LH_2 created during the fuelling process to be recovered. Furthermore, fuelling brooms alongside the extended air passenger bridge can also be considered as additional alternatives. Last but not the least, the system has to be safe, at least at the level of today's conventional Jet A fuel supply system [24].

6.3 Methodology for Assessing the Potential of Liquid Hydrogen

The methodology for assessing the potential of LH_2 as an alternative aviation fuel for greening the air transport system and particular airports consists of the following corresponding models: (1) The model for assessing the potential at the scale of the entire air transport system, and (2) The model for assessing the potential at the scale of an individual large airport.

6.3.1 Previous Research

In general, the research on the potential use of LH_2 as an ultimately unlimited and relatively cheap source of (air) transportation fuel instead of currently used fossil fuels has been continuously carried out over the past four decades. This research has particularly intensified after the oil crisis in 1970s. Much of it has been dealing explicitly and/or implicitly with a wide range of issues. In the scope of such scientific/professional efforts, Peschka and Wilhelm [29] have given one of the first comprehensive descriptions of LH_2 as an alternative fuel, its prospective applications of which one could certainly be in air and surface transportation, and some consequent environmental impacts. Specifically, a great deal of research has been devoted to assessing potential use of LH_2 for both commercial (civil) and military aviation. In general, this research can be broadly classified into three categories. The first category includes research relating to the characteristics of LH_2 as a prospective aviation fuel and its comparison to conventional, currently used aviation fuels. In addition, such research often deals with the prospective (re)-design of conventional and supersonic aircraft in order to allow them to be powered by LH_2 [6–9, 27, 28, 31, 32]. The second category of research deals with the overall logistics of LH_2 as an aviation fuel at commercial (civil) airports. Specifically, such research includes the design, installation, and operation of LH_2 fuel-supply systems at such airports, including elaboration of their technical/technological, operational, economic, and safety characteristics [3, 24, 25]. Finally, the last category includes the most recent research which has aimed to investigate if and by how much LH_2 could be a potentially environmentally friendlier transportation and particularly air transportation fuel compared to its fossil fuel counterparts. This has mainly included dealing with the characteristics of air pollution from burning LH_2 and their comparison with burning hydrocarbon fuels, and estimation of the contribution of LH_2 to the future reduction of greenhouse gas emissions from global commercial air transportation. In most cases, the above-mentioned issues have been addressed simultaneously but at differing levels of detail [4, 12]. In addition, the potential of LH_2 to stabilise and even diminish air pollution in the long term from the global-air transport system and local-airport scale has been demonstrated by Janic [22, 23].

6.3.2 Objectives and Assumptions

The main objectives are to develop a methodology, which will enable the following issues to be considered:

- Estimating the quantities of emissions of air pollutants at the level of the air transport system and in the airside area of an individual airport by LTO (Landing and Take-Off) cycles carried out according to given scenarios. In both cases, conventional Jet A fuel and cryogenic powered aircraft are considered to operate in different proportions over the observed period of time;
- In particular, this includes estimating fuel production and storage capacities at a given airport depending on the volume and structure of demand in terms of the number of flights and the quantity of fuel required to perform such flights during particular time intervals of the period concerned;

- Carrying out a sensitivity analysis of the quantities and structure of air pollutants with respect to changes in the most influential factors, such as the volume of traffic at the global-system and local-airport scale, the average aircraft size, and the proportion of cryogenic aircraft in the airport fleet mix during the period concerned; and
- Investigating the time horizon in which the system on the one hand and a given airport on the other, despite continuous traffic growth, could expect stagnation and even a decrease of air pollution, i.e. the time after which the air transport system as a whole and particularly a given airport can be considered as more sustainable, i.e. greener, in terms of air pollution.

In developing the methodology for assessing emissions of air pollutants by the air transport system and at a given airport's airside area, the following assumptions are made:

- Air traffic grows continuously at the system and airport scale in terms of the annual number of flights and LTO cycles, respectively, over time. In the case of a given airport, traffic grows up to the level of saturation of the airport runway system capacity;
- LH_2 is always available in the required quantities on both levels;
- The inherent characteristics (time, engine thrust settings, fuel consumption, and emission rates in particular phases) of flights and corresponding LTO cycles of a given (aircraft/flight) category are approximately constant in each year of the period concerned;
- The introduction of cryogenic aircraft is assumed to be a safe gradual process with an incremental increase in their proportion in each time interval (year) of the period concerned;
- The fuel consumption of cryogenic aircraft during the particular phases of LTO cycles is analogous to that of conventional aircraft modified for the specific difference(s) in the characteristics of particular fuel types; and
- The quantities of air pollutants at both the global-system and local-airport scale consist of only direct emissions generated by flights and corresponding LTO cycles, respectively. This assumption is not expected to compromise the results and conclusions.

6.3.3 Model for the Air Transport System

6.3.3.1 Basic Structure

A "carbon–neutral" air transport system implies that the volumes of air traffic demand will continue to grow over a given period of time while their contributions to fuel consumption and related emissions of the greenhouse gases will remain constant or even decrease. A model for estimating the global annual quantities of greenhouse gas emissions from the air transport system can have the following form:

$$E_n = V_0(1 + i_v)^n F_{C0}(1 - i_f)^n \sum_{l=1}^{L} e_l \quad (6.1)$$

where

E_n is the total emission of greenhouse gases in year (n) counted from the beginning of a given period of N years, i.e. the base year "0" (tons);

V_0 is the volume of air traffic demand in the base year (0) of a given period (RPK[3]);

F_{C0} is the average unit fuel consumption of conventional jet fuel in the base year (0) of a given period (g/RPK);

i_v is the average annual rate of growth of traffic demand in terms of equivalent RPKs over a given period of time (%);

i_f is the average annual rate of improvements of average unit fuel consumption over a given period (%); and

e_l is the emission rate of l-th greenhouse gas (g/g of Jet A fuel)

According to expression (6.1), total emissions E_n can be influenced by affecting the influencing variables in the given (target) year (n) as follows:

- Achieving a higher rate of improvement of average unit fuel consumption compared to rates of air traffic growth, i.e. $i_f \geq i_v/(1 + i_v)$;
- Slowing air traffic growth according to the rate of improvement in the unit fuel consumption, i.e. $i_v \leq i_f/(1 - i_f)$;
- Constraining air traffic growth by imposing a cap on the total emissions of greenhouse gases, i.e. $i_v = \left[E_n^*/[V_0 F_{C0}(1 - i_f)^n \sum_{l=1}^{L} e_l]\right]^{1/n} - 1$, where E_n^* is the "cap" on the total emissions of the greenhouse gases in the target year (n); and
- Affecting air traffic growth rate by weakening its relationship with the main internal and external demand-driving forces.

At present, the first three of the above conditions are not likely to be achieved before 2025/2026 and beyond, mainly because of the relatively wide difference between the current and predicted average annual air traffic growth rates (3.1%, [19]; 5.4%, [2, 5] and the rates of improvements in fuel efficiency (1.2–2.2%; [26]. For example, in the first case i_f should be not less than 4.3–4.8%, respectively, which is almost twice as much as the current very optimistic 2.2%. In the second case, the air traffic growth rate i_v should not be greater than the expected rate of improvements in fuel efficiency, i.e. about 1.2–2.2%. In the third case, the main problem appears to be the criteria for setting up the annual cap E_n^* and its

[3] Equivalent RPKs are regarded as the sum of RPKs and RTKs (Revenue Ton Kilometres) (1 RTK = 10 RPK).

monitoring and control [19, 20]. However, the latter case seems to be only a highly uncertain expectation.

Consequently, it appears that the only realistic but certainly not sufficient alternative remains the above-mentioned expected reduction in fuel consumption and related emissions by technological and operational improvements. This indicates that achieving a "carbon–neutral" air transport system will be extremely difficult if not impossible with conventional aircraft jet fuels.

6.3.3.2 The Model Structure for Two Aircraft Fuel Technologies

The process of introducing cryogenic aircraft powered by LH_2 is expected to imply the gradual replacement of part of conventional aircraft fleets. This process will be able to start when the following conditions are fulfilled:

- A pallet of different categories of cryogenic aircraft are fully developed regarding the size-range (small-short, medium-medium, large-long);
- Sufficient manufacturing capacities of cryogenic aircraft and LH_2 are available to satisfy a given rate of replacement;
- Airport infrastructure for supplying LH_2 is fully operational;
- Market prices of LH_2 are competitive to prices of conventional jet fuel; and
- Emissions of greenhouse gases during LH_2 production are captured and stored.

The gradual replacement process will take place over a "transition" period, during which both conventional and cryogenic aircraft will be used. The contribution of such a "hybrid" fleet to the total emissions of the greenhouse gases in the year (k) of the "transition" period of K years can be estimated based on expression (6.1), as follows:

$$E_k = V_0(1+i_v)^k \left[F_{C01}(1-i_f)^k(1-ki_h)\sum_{l=1}^{L} e_l + F_{C02}(ki_h)\sum_{m=1}^{M} e_m \right] \tag{6.2}$$

where

i_h is the average share of the total volume of traffic (RPKs) carried out by cryogenic aircraft in each year of the observed period ($0 \leq ki_h \leq 1$; $k = 1, 2, \ldots, K$);

F_{C01}, F_{C01} is the average unit fuel consumption of conventional (Jet A) and cryogen (LH_2) fuel, respectively, in the base year (0) of the given "transition" period (g/RPK);

e_m is the emission rate of m th greenhouse gas from cryogen fuel (LH_2) (g/g of Jet A fuel)

The other symbols are analogous to those in the expression (6.1).

In expression (6.2), the parameter E_{C01} is assumed to be at the level achieved when the process of introducing cryogenic aircraft starts, i.e. at the beginning of the "transition" period, and will continue to improve over this period. The parameter E_{C02} will be lower than E_{C01} approximately proportionally to the ratio between the specific energy of conventional jet fuel and LH_2, i.e. 43.2/120 = 0.36. This ratio is assumed to remain constant over the "transition" period. Cryogenic aircraft replacing conventional aircraft will be introduced each year in a constant proportion, implying their constantly increasing share in satisfying air traffic demand (RPKs).

In this case, the eventual stabilisation and/or even reduction in emissions of greenhouse gases in the given (target) year could be achieved through the same alternatives as in expression (6.1). In addition, one additional alternative could be adjusting the rate of introducing cryogenic aircraft in expression (6.2) as follows:

$$i_h = \left[i_v F_{C01}(1 - i_f) \sum_{l=1}^{L} e_l \right] / \left\{ \left[F_{C01}(1 - i_f) \sum_{l=1}^{L} e_l - F_{C02} \sum_{m=1}^{M} e_m \right] [1 + i_v(k + 1)] \right\}. \tag{6.3}$$

where all symbols are analogous to those in the previous expressions.

6.3.4 A Model for an Airport

The methodology consists of two sub-models: One dealing with estimating the capacity of fuel production and storage at a given airport; and the other enabling estimation of the quantity of emissions of air pollutants by LTO cycles during particular intervals of the period concerned.

6.3.4.1 Sub-Model for Determining the Quantity of Fuel Production and Storage Capacity

The sub-model for determining the quantity of fuel production and storage capacity implies estimating the demand for a given type of fuel during the specified period, usually one day. This requires specifying the daily number of flights and their structure in terms of duration (short haul, medium haul, and long haul), and quantity and type of fuel required. Consequently, the average quantity of fuel of type (*j*) for an average flight departing from the airport during the day of the (*m*)-th interval (year) of the period concerned can be estimated as follows:

$$\overline{FC_{m/j}} = \sum_{i=1}^{I_m} (1 - q_{ij})^m F_{0ij} p_{m/ij} t_{m/ij} \tag{6.4}$$

where

q_{ij} is the average annual rate of improvement in fuel efficiency of aircraft type (*i*) powered by fuel type (*j*);

F_{0ij} is the rate of consumption of fuel type (j) by aircraft type (i) during a flight carried out in the starting interval (year) of the period concerned (kg/h);
$p_{m/ij}$ is the proportion of aircraft/flights of type (i) using fuel type (j) during interval (m) of the period concerned;
$t_{m/ij}$ is the average duration of flights of type (i) using fuel type (j) during interval (m) of the period concerned; and
I_m is the number of different aircraft/flight types demanding fuel type (j) at a given airport in the (m)-th interval of the observed period

Based on expression (6.4), the daily demand (tons) for fuel type (j) at a given airport can be estimated as follows:

$$D_{m/j} = N_{m/j}\overline{FC_{m/j}} \quad (6.5)$$

where
N_{mj} is the number of flights carried out by aircraft using fuel type (j) in the year (m) of the period concerned

Specifically, if the fuel type (j) is LH_2, it will be produced on-site during liquefaction. If the available time for production is $T_{m/j}$ (usually during the eight night hours when most airports do not operate due to noise constraints), the installed production capacity (tons of LH_2/h) will amount to about:

$$PC_{m/j} = D_{m/j}//T_{m/j} \quad (6.6)$$

where the symbols are as in the previous expressions.

The quantities produced (or delivered) by the end of time $T_{m/j}$ will have to be stored in reservoirs, whose capacity (m^3) without additional spare space can be determined as follows:

$$V_{m/j} = D_{m/j}/v_j \quad (6.7)$$

where
v_j is the specific density of fuel type (j) (kg/m^3).

6.3.4.2 Sub-Model for Estimating Emissions of Greenhouse Gases

The basic structure of the sub-model for estimating emissions of air pollutants is based on: specifying the period concerned, i.e. the time horizon within which the estimation of quantities of emissions of air pollutants at a given airport is to be carried out; the number of ATMs, i.e. the LTO cycles in each interval (year) of the observed period; the structure of LTO cycles in terms of average aircraft size and type of fuel used; the time and rate of introducing cryogenic aircraft; and the rate of consumption of particular fuel types and the corresponding rates of emissions of particular air pollutants during particular phases of LTO cycle(s).

Consequently, the total emissions of the (*l*)-th type of air pollutant during the LTO cycle carried out by an aircraft of type (*i*) using fuel type (*j*) in the (*m*)-th year of the period concerned can be estimated as follows:

$$Q_{mijl} = N_{mij}E_{mijl} = N_{mij}\sum_{k=1}^{K_{ij}}(1-q_{ij})^{m}F_{0ijk}t_{mijk}e_{mijkl} \tag{6.8}$$

where

- N_{mij} is the number of LTO cycles carried out by aircraft type (*i*) using fuel type (*j*) in the interval (year) (*m*) of the period concerned;
- E_{mijl} is the quantity of air pollutant (*l*) emitted during the LTO cycle carried out by aircraft type (*i*) using fuel type (*j*) in the interval (year) (*m*) of the period concerned (kg);
- F_{0ijk} is the rate of consumption of fuel type (*j*) by aircraft type (*i*) during the (*k*)-th phase of the LTO cycle carried out in the starting interval (year) of the period concerned (kg/h);
- t_{mijk} is duration of the (*k*)-th phase of the LTO cycle carried out by aircraft type (*i*) using fuel type (*j*) in the (*m*)-th interval (year) of the period concerned (h);
- e_{mijkl} is the emission rate of the (*l*)-th air pollutant by aircraft type (*i*) using fuel type (*j*) during the (*k*)-th phase of the LTO cycle carried out in the (*m*)-th interval (year) of the period concerned (kg/kg of fuel); and
- K_{ij} is the number of phases of the LTO cycles carried out by aircraft type (*i*) using fuel type (*j*)

In expression (6.7), the number of LTO cycles N_{mij} can be determined as follows:

$$N_{mij} = N_{m}p_{m/ij} \tag{6.9}$$

where:

- N_m is the total number of LTO cycles carried out in the interval (year) (*m*) of the period concerned

In expression 6.8, the following condition should be satisfied:

$$\sum_{i=1}^{I_m}\sum_{j=1}^{J_m}p_{m/ij} = 1 \tag{6.10}$$

where

- I_m is the total number of LTO cycles carried out by an aircraft type (*i*) in the year (*m*) of the period concerned; and

J_m is the number of different fuel types used in the interval (year) (m) of the period concerned

From expression (6.7), the total emissions of air pollutant (l) during the LTO cycle carried out in the interval (year) (m) of the period concerned can be estimated as follows:

$$Q_{ml} = \sum_{i=1}^{I_m} \sum_{j=1}^{J_m} Q_{mijl} \tag{6.11}$$

where all symbols are as in the previous equations.

Summing up the annual emissions calculated from Eq. 6.11 enables estimation of the total emissions of greenhouse gas (l) over the given period.

6.4 Application of Proposed Methodology

6.4.1 Inputs–the Model for the Air Transport System

The proposed model for estimating the potential of LH_2 for making the air transport system more sustainable, i.e. "greener" in terms of air pollution is applied to the long-term development of the air transport system and the related emissions of greenhouse gases. The time horizon is divided into three sub-periods: 2006–2025, 2026–2040, and 2041–2065.

The first sub-period is specified by the two leading aircraft manufacturers [2, 5]. The second sub-period is specified as the period before the start of large scale introduction of cryogenic aircraft (the year 2041), while the final period represents the "transition" period of gradually replacing a certain proportion of conventional aircraft with cryogenic aircraft. This implies that at the end of the final period, a "hybrid" aircraft fleet consisting of both aircraft categories will be in operation. The inputs for the model characterising each sub-period are given in Table 6.3.

As can be seen, the growth rates of air traffic demand are assumed to be constant during each sub-period, but decrease when looking further into the future.[4] This reflects the increasing maturity of the air transport market (demand) combined with the weakening dependency between air transport demand and its

[4] The average growth rate of air traffic demand over the entire time horizon is about 3.2%, which is similar to the growth rate of 3.1% over the period 1990–2050 in one of the scenarios of traffic growth developed by IPCCs. This rate produces a total of about 16.5 trillion RPKs in the year 2050 and 26.02 trillion RTKs in the year 2065 [19].

Table 6.3 The input data for scenarios of influencing cryogenic aircraft

Input variable	Period		
	2006–2025	2026–2040	2041–2065
Basic annual traffic volume: V_0 (trillion Equivalent RPKs)	6.26[a]	13.78	22.61
Average traffic growth rate: i_v (%)	5.4[a]	3.5	2.0
The number of aircraft at the beginning of the period	18,230[a]	36,420	48,823
Average aircraft utilisation an at the beginning of the period (trillion RPKs/yr)	0.362	0.378	0.463
Rate of improvement in aircraft utilisation: (%/yr)	1.50	1.25	1.00
Average unit fuel consumption of conventional aircraft: E_{C01} (g/RPK)	27.70	19.66	16.28
Rate of improvement in E_{C01}: i_f (%/yr)	1.70	1.25	1.00
Average unit fuel consumption of cryogenic aircraft: E_{C02} (g/RPK)	–	–	5.86[b]
Average share of total traffic carried by cryogenic aircraft: i_h (%/yr)	0.00	0.00	1.00/2.00

[a] Compiled from [2, 5]
[b] $E_{C02} = 0.36\ E_{C01}$

main driving forces. Improvements in unit fuel consumption of conventional aircraft are assumed to be permanent, albeit occurring at a decreasing rate over time. In addition, aircraft utilisation is assumed to generally increase over time at a decreasing rate, which also implies increasing the number of aircraft at a decreasing rate. The rates of introducing cryogenic aircraft are assumed to be constant in each year of the "transition" period, giving cryogenic aircraft a share of 22 and 50% of RPKs by the end of 2065. Eventual improvements in unit fuel consumption of cryogenic aircraft are not considered due to the lack of realistic data.

6.4.2 Inputs–the Model for an Airport

The proposed model for estimating the potential contribution of LH_2 to the more sustainable development, i.e. greening of airports in terms of air pollution is applied to one of the world's largest airports–London Heathrow Airport (UK)–using "what-if" reasoning. For this purpose, two scenarios of airport demand development are elaborated in terms of: the annual number of atms (air transport movements—1 atm corresponds to 1 landing or 1 take-off), and the change in the structure and fuel efficiency of the aircraft fleet, in relation to aircraft size and the rate of introducing cryogenic aircraft on the one hand and solutions for providing airport runway capacity in the long term, i.e. from the year 2010 to the year 2065, on the other.

6.4.2.1 Demand

The Volumes

Over the past two decades (1991–2008), traffic in terms of the annual number of ATMs, as an indicator of the annual number of LTO cycles, has grown at an average annual rate of +1.72%. Growth was faster during the first half than during the second half of the period; in the first half, market forces mainly drove the number of atm. In the second half, environmental constraints in terms of noise and air pollution were strengthened resulting in capping the maximum number of available slots. In addition to other effects, this has created slot scarcity and stimulated the use of larger aircraft. Consequently, growth began to stagnate during the last few years of the period when demand closely approached the available airport runway capacity, which, in turn continued to stimulate an even greater increase in aircraft size in order to use the scarce slots as efficiently as possible.

The forthcoming period is expected to be twice as long as the previous one, i.e. from 2010 to 2065. Such a long period had been selected due to expectations that during its second half, i.e. 2040–2065, cryogenic aircraft will be gradually introduced. As a result, during this sub-period, both cryogenic and conventional jet-fuel aircraft are expected to operate simultaneously at airports.

The character of growth of the annual number of atm or LTO cycles over this long period is expected to be generally similar to that in the previous period. This implies the following:

- The annual number of atm will never exceed the available airport runway capacity;
- By the end of the period concerned, the annual number of atm will again approach available airport capacity at that time; and
- During the observed period, different forces are expected to drive airport demand at different growth rates, generally either positively or negatively influencing demand (Janic 2009).

Consequently, the following growth rates of ATMs during the observed period are given in Table 6.4.

As can be seen, the annual growth rates of atm are assumed to generally decrease over time, reflecting the gradual saturation of airport runway capacity and further increases in aircraft size. Under such circumstances, larger conventional aircraft are assumed to be more likely replaced by cryogenic aircraft.

Consequently, according to the above-mentioned growth rates, the annual number of atm is expected to increase to about 479,000 in 2015, 534,000 in 2020, 540,000 in 2025, 639,000 in 2030, 696,000 in 2035, and 736,000 in 2040 and beyond, until the end of the observed period. The latter figure is also the limit specified by the runway system operational capacity described below.

Table 6.4 Growth rates of atm in the given example

Period	Growth rate of atm (%/year)
2010–2020	2.2
2020–2035	1.8
2035–2050	1.4
2050–2065	0.8

The Structure

The aircraft fleet structure has and will continue to consist of four aircraft/flight categories: Medium aircraft with an average flight distance of 700 km (A319/320 s/B737 s); Large 1 aircraft with an average flight distance of 5,500 km (A330/B767/777); Large 2 aircraft with an average flight distance of 7,500 km (A340/B747); and Large 3 aircraft with an average non-stop flight distance of 9,000 km (B747/A380).

During the period 2010–2040, the fleet structure is assumed to change as follows: the proportion of medium aircraft/flights is expected to decrease from the current proportion of 65% at an annual rate of 2.5% per year. This will also occur due to the fact that the airport will be connected to the High Speed Rail network, as well as the fact that airlines intend to use the scarce and increasingly expensive slots more efficiently. The proportion of Large 1, Large 2, and Large 3 aircraft is expected to increase by 1% per year. During the period 2040–2065, the fleet structure is assumed to be constant as follows: Medium–50%, Large 1–30%, Large 2–16%, and Large 3–4%. In such contexts, one Medium aircraft is assumed to require 4.4, Large 1–57, Large 2–104, and Large 3–153 t of Jet A fuel per departure.

Table 6.5 summarises the above-mentioned characteristics.

In addition, during the period 2040–2065, Large conventional aircraft will gradually be replaced with cryogenic counterparts at an annual rate of 1.5–2.0%., which will also contribute to the changing structure of LTO cycles.

Fuel Consumption During the LTO Cycle

Estimation of fuel consumption during LTO cycles of particular types is carried out after the aggregation of diverse aircraft types into particular categories. Therefore, as in the case of fuel demand for the entire flight, fuel consumption per LTO cycle refers to an average aircraft, despite differences between aircraft of the same type operating between the airport and different regions/countries (see Table 6.5). This implies that the time spent in particular phases of the LTO cycle and the engine-thrust setting for both Medium and Heavy conventional and cryogenic aircraft are as follows: take-off 0.7 min, with 100% thrust, climb 2.2 min with 85% of thrust, approach 4.0 min with 30% thrust, and taxi-idle 26 min with 7% thrust [17].

Table 6.5 Characteristics of aircraft fleet categories in the given example

A/C- Flight category	A/C type	A/C TOW[a] (t)	A/C Capacity (seats)	Route length (km)	Flight duration (h)	Fuel/flight[b] (tons-Jet A)
S-M[c]	A320s/ B737	77	159	700	1.8	4.4
LR1[d]	A330/ B777	284	357	5,500	6.7	57.0
LR2[e]	A340/ B747	398	458	7,500	8.8	104
LR3[f]	A380/ B747	495	592	9,000	10.5	153

[a] Take-Off Weight
[b] The equivalent amount of LH_2 is about 0.36 of the specified amount of Jet A fuel
[c] Short-Medium
[d] Long Range 1
[e] Long Range 2
[f] Long Range 3

Table 6.6 Rates of improvement of fuel efficiency of conventional aircraft fleet

Period	Rate (%)	
	Entire flight	LTO cycle
2010–2025	1.70	0.75
2025–2040	1.50	0.50
2040–2065	1.00	0.25

Compiled from: [22]

In addition, the calculation implicitly takes into account improvements in aircraft fuel efficiency, assumed to be achieved for each specific aircraft category (type). The average annual rates of these improvements are given in Table 6.6.

The above-mentioned differences between rates of fuel efficiency improvement for the entire flight and the LTO cycle are assumed to occur mainly due to increasing congestion at the given airport, which reduces improvements that could be achieved for the entire flight. Consequently, the corresponding average fuel consumption per LTO cycle is estimated to be: 0.99 t of conventional Jet A fuel for Medium, 2.67 t for Large 1, 3.6 t for Large 2, and 4.4 t for Large 3 aircraft. Regarding differences in specific energy, the consumption of LH_2 during the LTO cycles will amount to approximately only 36% of the above-mentioned equivalent amount of Jet A fuel (kerosene) (see Table 6.1).

Emission Rates

The emission rates of particular greenhouse gases are applied from Table 6.1 as follows. For Jet A fuel: 3.15 kg of CO_2/kg of fuel, 1.23 kg of H_2O/kg of fuel, 0.41 g of NO_x/kg of fuel, and 0.84 g of SO_4/kg of fuel; for LH_2: 0.00 kg of CO_2/kg of fuel, 3.20 kg of H_2O/kg of fuel, 0.105 g of NO_x/kg of fuel, and 0.00 g of SO_4/kg of fuel [21].

6.4.2.2 Capacity

Airport runway capacity has proved to be the "critical" component of the capacity of the entire airport system. Due to environmental (noise and air pollution) burdens, a cap of 480 thousand atm/year has been set up [13]. Since annual demand has almost reached this capacity, most recently, the UK Government, after a long public inquiry, has approved two solutions for gradually increasing this capacity over the period concerned [13].

Solution 1 called "*changing the runway operating mode*" includes changing the operating mode of the existing two–runway system from "segregated" to "mixed" mode. The former implies using each runway exclusively for landings or take-offs, while the latter implies using both runways simultaneously for both landings and take-offs. Such change will provide a total hourly capacity of the two-runway system of 96 atm/h (2×48 atm/h = 96 atm/h), and an annual capacity of 560,000 atm/yr, which is about 17% higher than the present cap of 480,000 atm/year (2×48 atm/h $\times$ 16 h/day $\times$ 365 d/year = 540,640 atm/y). This *Solution* is expected to be implemented by the year 2015 [13].

Solution 2 called "*building a new runway*" implies building a new third parallel runway north of the existing two runways (as shown in Fig. 4.7) (BAA 2005) [13]. This alternative was supported by the former UK labour Government in power until the middle of 2010. The newly elected conservative-liberal coalition Government has started to prioritise developing the airport into a true multimodal hub as described in Chap. 4 (see Fig. 4.7).

The new runway will operate in the "mixed" mode while the existing two parallel runways will once again operate in the "segregated" mode for 16 h/day, as in *Solution* 1. Such a three-runway system will provide an hourly capacity of 126 atm/h (78 + 48 atm/h) and an annual capacity of 736,000 atm/year (126 atm/h $\times$ 16 h/day $\times$ 365 days/year = 735,840 atm/yr). The annual capacity of the three runway system will be gradually increased starting from 605,000 atm/year to 702,000 atm/year by 2035, and 736,000 atm/year by 2045 and beyond, until the end of the observed period (2065) [13]. It will be possible to achieve such an increase in capacity due to the introduction of more noise[5] and air pollution efficient aircraft, the latter particularly thanks to replacing part of the conventional aircraft fleet with cryogenic aircraft.

---

[5] For example, this implies that the area of land with the noise contour of 57 dBA will decrease to around 113 km and the exposed population to 206,000 (currently, it is 252,000). In addition, the concentration of NO_x is expected to continue to be within the above-mentioned tolerable limits [13].

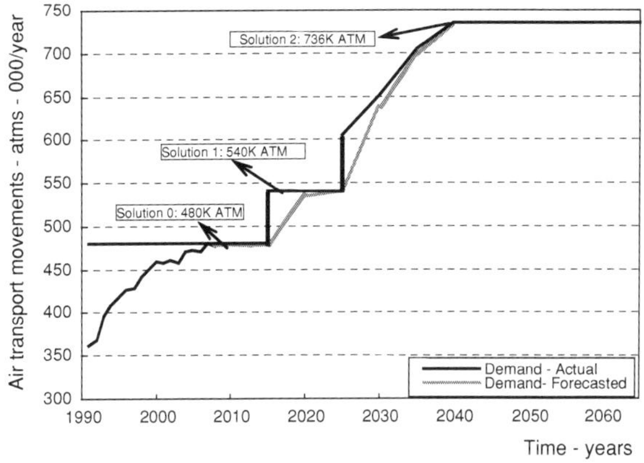

Fig. 6.1 Matching capacity to demand in the long-term in the given example (Compiled from [13])

6.4.2.3 Matching Airport Runway Capacity to Demand

Figure 6.1 shows possible scenarios of developing demand in terms of the annual number of atm, implicitly but simultaneously influenced by the above-mentioned factors, i.e. the driving forces, and its matching with the expected development of the runway system capacity.

6.4.3 Results for the Air Transport System

The results of applying the model for the air transport system using the above-mentioned inputs in Table 6.3 are shown in Fig. 6.2a, b, c. This figure shows the development of air transport demand and related emissions of greenhouse gases dependent on time in relative terms (Index). The base year of 2006 is adopted. Specifically, Fig. 6.2a shows the development of emissions of CO_2, Fig. 6.2b of H_2O, and Fig. 6.2c of NO_x.

Figure 6.2a shows that if only conventional aircraft continue to be used in the future, emissions of CO_2 will continue to increase in line with increased air traffic volumes. However, emissions of CO_2 will rise more slowly than the traffic, mainly due to permanent improvements in aircraft unit fuel consumption on the one hand and aircraft utilisation on the other. For example, at the end of the period (the year 2065), air traffic will increase six fold and the related emissions of CO_2 3.5 fold compared to the base year 2006. This is lower than in the IPCC's Reference Scenario where CO_2 emissions in the year 2050 are predicted to be about 3.9 times greater than in the year 2006 [19]. Consequently, it becomes evident that independent of the rate of improvement of conventional aircraft, stabilisation of global annual CO_2 emissions will not be possible under conditions of unconstrained traffic growth, and thus achieving a "carbon neutral" system seems unlikely. However, from the moment when cryogenic aircraft are introduced even in a modest proportion of only about 1% per year, despite continuous traffic growth, emissions of CO_2 start to gradually slow down, stagnate, and finally stabilise in the

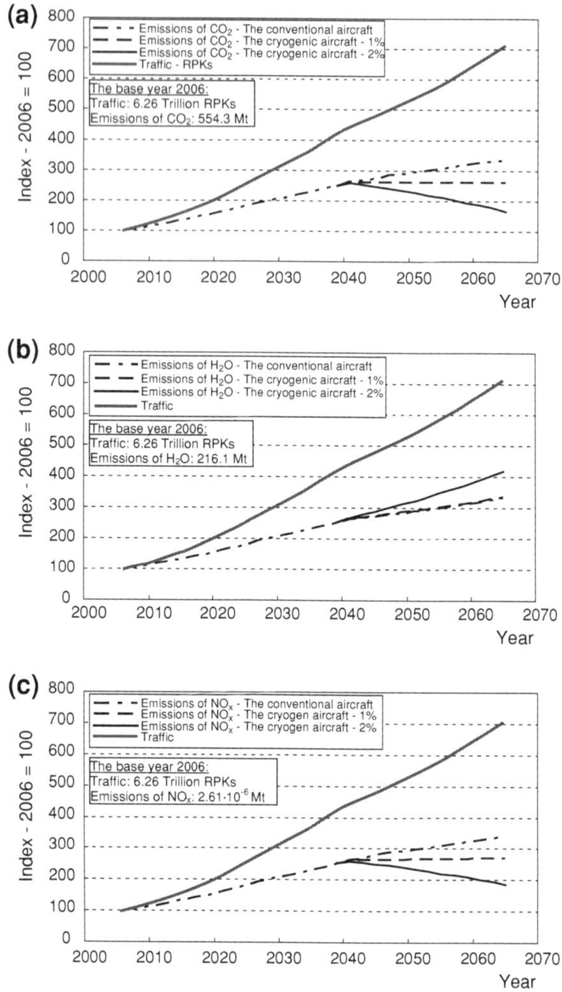

Fig. 6.2 The prospective influence of cryogenic aircraft on long-term global emissions of greenhouse gases. (Compiled from [22]) **a** CO_2 emissions. **b** H_2O emissions. **c** NO_x emissions

year 2065 at a level about 2.8 times higher than in 2006. If the rate of introduction of cryogenic aircraft is increased to about 2% per year, the CO_2 increase rate will immediately start to decrease to about 1.8 times higher in the year 2065 compared to the base year 2006. This shows that cryogenic aircraft may enable decoupling of the growth of air traffic and related emissions of CO_2 and thus contribute to achieving a "carbon neutral" air transport system.

Figure 6.2b shows that H_2O emissions will continue to increase with air traffic volumes independent of the technology in use. If only conventional aircraft are used, the levels of H_2O in 2065 will be about 3.3 times greater than in the base year (2006). At the same time, air traffic volumes will be about 7 times higher. This indicates that, as in the case of CO_2, reducing unit fuel consumption and improving aircraft fleet utilisation can slow the increase in H_2O emissions.

Introducing a relatively low proportion of cryogenic aircraft (1%) will slightly (negligibly) increase this level during the period of replacement (2040–2065). However, if the proportion of cryogenic aircraft introduced is 2%, the level of H_2O in 2065 will be about 4.2 times higher than in the base year (2006). These figures confirm the present concerns that cryogenic aircraft will not stabilise the H_2O emissions but rather the opposite—they will contribute to their substantial rise and thus to the increased risk of more intensive contrail formation.

Figure 6.2c shows the prospective long-term emissions of NO_x. As can be seen in the case of the other two greenhouse gases, when conventional aircraft are exclusively used over the entire period (2006–2065), NO_x emissions continue to rise due to the growth of traffic, but again at a slower rate, mainly thanks to improvements in unit fuel consumption and aircraft utilisation. This again indicates that conventional aircraft will not be able to stabilise the level of NO_x and thus make the system neutral under conditions of unconstrained traffic growth. For example, the level of NO_x in 2065 will be about 3.5 times greater than in the base year (2006), compared with an approximately 7 fold increase in air traffic volumes. If cryogenic aircraft really achieve emission rates of NO_x of about 5–25% of the rate of conventional aircraft, their gradual introduction, depending on the rate, will certainly stabilise and even decrease the level of NO_x, despite air traffic growth. For example, if the rate of introduction of cryogenic aircraft is 1%, emissions of NO_x in 2065 will stabilise at a level of about 2.8 times the level in the base year (2006). If the rate of introducing cryogenic aircraft is 2%, emissions of NO_x will decrease by 2065 to the level of about 2 times higher than in the base year (2006).

6.4.4 Results for the Selected Airport

The results from the calculations of the fuel demand (consumption) and the emissions of air pollutants at the given airport are shown in Figs. 6.3, 6.4, 6.5 and 6.6a, b, c.

Figure 6.3 shows that the average daily consumption (demand) for conventional Jet A fuel first increases and then stabilises in line with the increasing and/or decreasing volume of atm during the observed period.

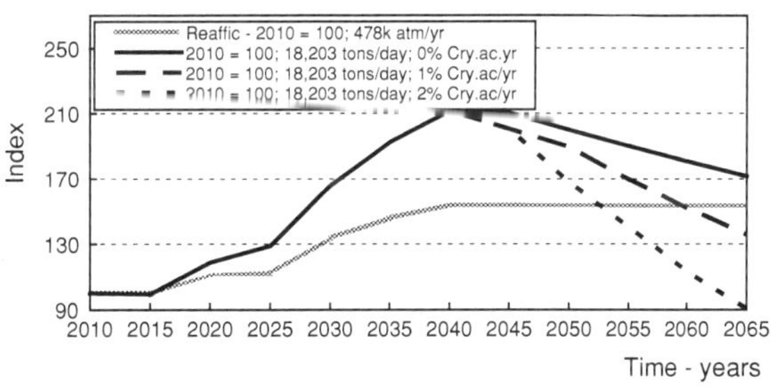

Fig. 6.3 The average daily demand for Jet A fuel depending on the annual number of atm and rates of introducing cryogenic aircraft during the observed period in the given example

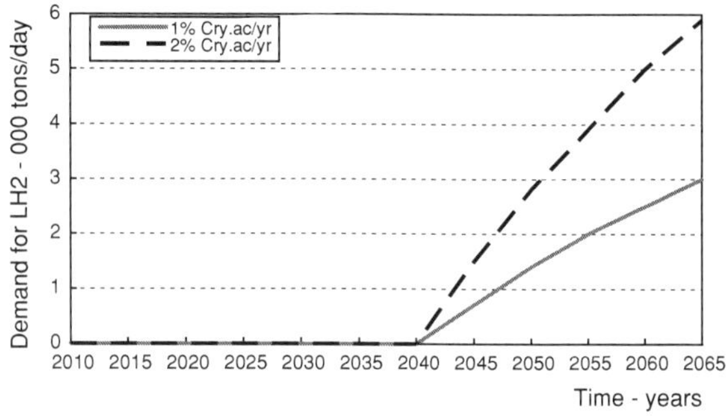

Fig. 6.4 The average daily demand for LH_2 fuel, depending on the rates of introducing cryogenic aircraft during the observed period in the given example

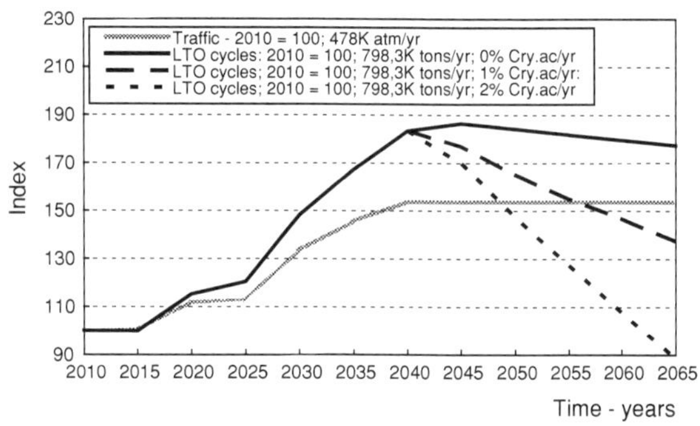

Fig. 6.5 Consumption of Jet A fuel depending on the annual number of LTO cycles and rates of introducing cryogenic aircraft in the given example

This increase is at a higher rate than the present rate of increase due to the gradual changing of the fleet structure in favour of increasing the average aircraft size despite continuous improvements in aircraft/flight fuel efficiency. However, when the volume of atm reaches the airport capacity, the demand for fuel stabilises and starts to stagnate and even slightly decrease, thanks to the continuing above-mentioned improvements in aircraft/flight fuel efficiency. Since the time of starting introduction of cryogenic fleet, which is expected to occur as airport capacity reaches saturation and the daily number of atm becomes constant, the demand for conventional fuel starts to decrease at an average annual rate approximately equivalent to the annual rate of introducing cryogenic aircraft.

Constrained traffic growth due to saturation of airport capacity, combined with technological improvements in fuel efficiency of conventional aircraft on the one hand, and gradual replacement of these with cryogenic aircraft on the other, reduces the daily demand for conventional fuel at the airport at a faster rate. At the same time, the demand for cryogenic fuel (LH_2) will continuously increase over time and with the increase in the annual rate of replacement of conventional aircraft. Figure 6.4 shows this development.

Consequently, from the begining of introducing cryogenic aircraft (2040), two fuel systems will have to operate at the airport simultaneously, one with decreasing

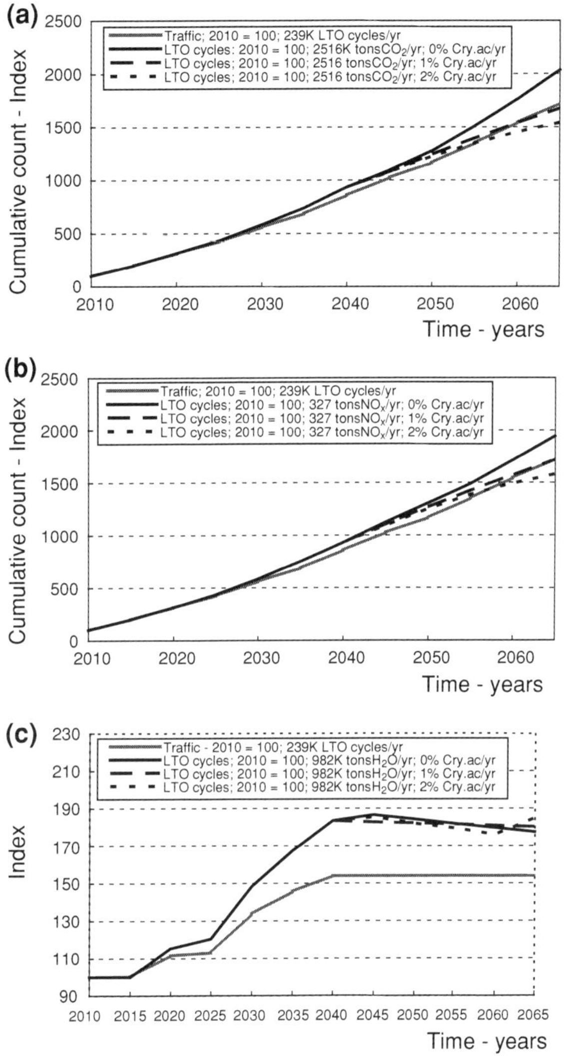

Fig. 6.6 Emissions of air pollutants depending on the annual number of LTO cycles and rates of introducing cryogenic aircraft in the given example. **a** Carbon Dioxide-CO_2. **b** Nitrogen Oxides-NO_x. **c** Water vapour

capacity and the other with increasing capacity during the observed period (2040–2065). At the beginning of that period, the demanded quantities of LH_2 fuel will be stored in 4–6 steel cylindrical tanks, each with a capacity of about 850 m^3. The number of these tanks will increase in line with the demand for LH_2 fuel to about 15–30 tanks by 2065 [24].

Figure 6.5 shows consumption of conventional Jet A fuel during the LTO cycles depending on the daily volume of atm during the observed period.

As can be seen, the effect is similar as in the case of fuel demand for the entire flight (see Tables 6.4 and 6.5). This implies that the overall consumption of conventional fuel during the LTO cycles will increase in line with the volume of traffic and the average aircraft size, despite simultaneous improvements in aircraft fuel efficiency. Introducing cryogenic aircraft (from 2040 on) will contribute to a decrease in the total fuel consumption during the LTO cycles in line with the proportion of such aircraft in the fleet. For example, if the rate of introducing these aircraft is 2%, total fuel consumption will drop to about 90% of the consumption at the beginning of the observed period (2010).

Figure 6.6a, b, c shows the emissions of air pollutants such as CO_2, NO_x, and H_2O, respectively, during the LTO cycles over the observed period depending on the volumes of ATMs in the given example. Specifically, Fig. 6.6a, b shows the cumulative emitted quantities of CO_2 and NO_x, respectively, from the beginning to the particular interval of the observed period. This way of presenting the effects corresponds to the time in which man-made CO_2 and NO_x emissions remain in the atmosphere; this amounts to about 90–200 and 120 years, respectively. As can be seen, the cumulative amounts of the two air pollutants increase in line with traffic volumes before and after their stagnation.

This implies that, even when traffic growth ends, cumulative emissions of air pollutants will increase, albeit at a different rate. In addition, the growth rate of both air pollutants over the longer parts of the observed period will be always higher than the growth rate of airport traffic if cryogenic aircraft fleet is not introduced, mainly because of increases in the average aircraft size and despite improvements in aircraft technology, i.e. in fuel consumption during the LTO cycles. With the introduction of cryogenic aircraft, the cumulative growth of two of the air pollutants will slow down, i.e., grow at a decreasing rate, particularly towards the end of the observed period. Such development implies that if cryogenic aircraft were introduced in a substantive proportion, this would appear to contribute to stabilising and even decreasing (i.e. compromising) the cumulative growth of the two air pollutants, CO_2 and NO_x, in the given example.

Figure 6.6c shows the development emissions of H_2O during particular intervals (years) of the observed period. This way of presenting such development is chosen due to the fact that water vapour (H_2O) remains near the surface for only a few days, although it is considered as one of the most significant greenhouse gases, both natural and man-made. As can be seen, if only conventional aircraft continued to operate at the airport during the observed period, the emissions of H_2O would increase in line with the number of LTO cycles and the average aircraft size, despite improvements in aircraft fuel efficiency. The latter will bring a slight decrease in these emissions under conditions of stagnation in the volume of LTO cycles and stabilisation of the fleet structure in terms of average aircraft size. The introduction of cryogenic aircraft under conditions of stagnating LTO cycle volumes of (2041–2065) will not cause a substantive stagnation of these emissions, as they will increase after the proportion of cryogenic aircraft is introduced at an annual rate of 2% increases more substantially, i.e., above 40% after 2060.

6.5 Concluding Remarks

This chapter has described the potential of cryogenic aircraft using LH_2 for achieving a "carbon neutral", i.e. "greener" global—air transport system and airports as its components in the long-term future. The elaboration based on "what-if" reasoning has resulted in developing a methodology consisting of dedicated models and their application to the appropriate cases. The results obtained from the model applied to the entire air transport system have suggested the following:

Global air traffic will continue to grow, driven by the main external and internal demand-driving forces, which implies its unconstrained growth during the given time horizon;

If aircraft powered by conventional jet fuel continue to be exclusively operated over the given time horizon, emissions of the main greenhouse gases such as CO_2, H_2O, and NO_x will continue to grow, despite continuous improvements in aircraft fuel efficiency and utilisation; this indicates that the system will not become "carbon–neutral" during the relevant time-horizon (2006–2065);

The gradual replacement of conventional aircraft with cryogenic aircraft could contribute to stabilising and even decreasing direct emissions of CO_2 and NO_x, despite the continued air traffic growth, thus creating conditions for a "carbon neutral" (air transport) system;

The substantively increased emissions of water (H_2O) by cryogenic aircraft and their much greater contribution to the formation of contrails remains the main matter of concern when considering the prospective overall benefits of LH_2 to the "carbon–neutral" air transport system.

The results obtained from the model developed and applied to an airport as a component of the air transport system (London Heathrow, London, UK) have suggested the following:

- For the period 2010–2065, the demand for conventional Jet A fuel at the airport is expected to generally increase, mainly driven by airport traffic growth in terms of ATM (Air Transport Movements) and average aircraft size, despite improvements in aircraft fuel efficiency. It will decrease considerably after the stagnation of airport traffic growth and the introduction of cryogenic aircraft. Similar trends apply to the LTO cycle(s). In parallel with decreased demand for Jet A fuel, demand for LH_2 fuel will increase. This will require providing storage capacities for both fuels, and their timely adaptation to changes in the corresponding demand;
- The cumulative and relative emissions of air pollutants such as CO_2, NO_x, and H_2O during the LTO cycles will generally increase, mainly driven by traffic growth and an increase in aircraft size without significant compensation from improvements of aircraft fuel efficiency. When traffic growth stagnates, the cumulative emissions of CO_2 and NO_x will continue to increase, despite improvements of aircraft fuel efficiency. When cryogenic aircraft is introduced, the rate of cumulative increase of these air pollutants will decrease, thus

indicating LH_2 as an alternative for their stabilisation and even possible decrease. However, the increased proportion of cryogenic aircraft will contribute to increase emissions of water vapour (H_2O) despite the stagnation of traffic growth and simultaneous improvements in aircraft fuel efficiency.

Consequently, at both the global-system and local- airport scale, LH_2 remains to be seriously considered as an alternative aviation fuel, which offers the potential for their simultaneous long-term sustainable, i.e. greener, growth with at least stabilised and then gradually decreasing cumulative emissions of greenhouse gases CO_2 and NO_x.

Cryogenic commercial aircraft powered by LH_2, related manufacturing plants for both vehicles and fuel, and airport fuel supply infrastructure do not exist yet. From the present perspective, switching from conventional Jet A fuel (kerosene) to LH_2 seems to be financially and technologically risky. In particular, the fact that the air transport system's share in the total man-made emissions of greenhouse gases is expected to range between 3 and 5%, increases these risks. The air transport system has, however, already undergone some radical changes. This prospective change is particularly challenging because it could enable the air transport system to act as an ultimate leader within the transport sector in creating a "carbon–neutral", i.e. "greener" environment for future mankind.

References

1. ACI (2008) Airports—actions on climate change. Airport Council International, Montreal, Canada
2. AIRBUS (2006) Airbus global market forecast. Airbus industries, Toulouse, France
3. Alder HP (1987) Hydrogen in air transportation feasibility study for zurich airport switzerland. Int J Hydrogen Energy 8(12):571–585
4. Behrend ES, Pohl HW, Roster N (1997) Hydrogen aircraft and airport safety. Renew Sustain Energy Rev 4(1):239–269
5. Boeing (2007) Current market outlook 2007: how will you travel through life? Boeing commercial airplanes: market analysis, Seattle, WA, USA
6. Brewer GD (1976) Aviation usage of liquid hydrogen fuel—prospects and problems. Int J Hydrogen Energy 1(1):65–88
7. Brewer GD (1978) Hydrogen usage in air transportation. Int J Hydrogen Energy 2(3):217–229
8. Brewer GD (1979) A plan for active development of LH_2 for use in aircraft. Int J Hydrogen Energy 3(4):169–177
9. Brewer GD (1982) The prospects for liquid hydrogen fueled aircraft. Int J Hydrogen Energy 1(7):21–41
10. Brewer GD (1991) Hydrogen aircraft technology. CRC Press, Boca Raton, FL
11. Chevron (2006) Alternative jet fuels addendum 1 to aviation fuels, technical reviews (FTR-3/A1). Chevron Corporation, USA
12. Coenen RM (2009) A proposal to convert air transport to clean hydrogen (CATCH). Int J Hydrogen Energy 19(34):8451–8453
13. DfT (2009) Adding capacity at heathrow airport: decisions following consultation. Report, department for transport, London, UK

14. EC (2000) The way to sustainable mobility: cutting the external cost of transport, brochure of the european commission, Brussels, Belgium
15. EC (2003) Liquid hydrogen fuelled aircraft system analysis (CRYOPLANE). European commission 5th R&D framework program (Growth 1998-2002). Brussels, Belgium
16. EWG (2007) Crude oil supply outlook. Energy watch group, report no. 3/2007, Berlin, Germany, p 101wRe
17. ICAO (1995) Aircraft engine emissions, environmental protection, annex 16, vol 2. International civil aviation organization, Montreal, Canada
18. IEA (2006) Hydrogen production and storage: research & development priorities and gaps. International energy agency, Paris, France
19. IPCC (1999) Aviation and the global atmosphere. Intergovernmental panel on climate change. Cambridge University Press, Cambridge
20. IPCC (2001) Climate change 2001–synthesis report. Contribution of working groups I, II, and II and III to the 3rd assessment report of IPCC, intergovernmental panel of climate change. Cambridge University Press, Cambridge
21. Janic M (2007) The sustainability of air transportation: a quantitative analysis and assessment, Ashgate, Aldershot
22. Janic M (2008) The potential of liquid hydrogen for the future 'carbon neutral' air transport system. Transp Res D 8(13):428–435
23. Janic M (2010) Is liquid hydrogen a solution for mitigating air pollution by airports? Int J Liq Hydrogen 5(35):2190–2202
24. Korycinski PF (1978) Air terminals and liquid hydrogen commercial air transports. Int J Hydrogen Energy 2(3):231–250
25. Korycinski PF (1978) Some early perspectives on ground requirements of liquid hydrogen air transports. Int J Hydrogen Energy 3(3):335–346
26. Learmount D (2007) New-technology aircraft to reduce average fuel consumption. (http://www.flightglobal.com)
27. Mikolowsky WT, Noggle LW (1978) The potential of liquid hydrogen as a military aircraft fuel. Int J Hydrogen Energy 4(3):449–460
28. Price RO (1991) Liquid Hydrogen—An Alternative Aviation Fuel? Int J Hydrogen Energy 16(8):557–562
29. Peschka W, Wilhelm LA (2001) Liquid Hydrogen: Fuel Of The Future, Springer Verlag, Berlin, Germany, p. 303
30. Typolev AA (1994) “”Utilization of Liquid Hydrogen ir Natural Gas as Aviation Fuels”, Conference Proceedings – Projects Energy '93, Ed. R. E. Billings and E. Dayton, International Academy of Sciences, Independence, Michigan, USA, p. 104
31. Wilkinson KG (1983) An airline view of LH2 as a fuel for commercial aircraft. Int J Hydrogen Energy 10(8):793–796
32. Winter CJ (1990) Hydrogen in high-speed air transportation. Int J Hydrogen Energy 8(15):579–595
33. Wood HJ, Long GR, Morehouse DF (2003) Long-term world oil supply scenarios: the future is neither as bleak or rosy as some asserts. EIA-energy information administration, Washington DC, USA. http://www.eia.doe.gov

Chapter 7
Greening the Airport Landside Area: Light Rail Rapid Transit Access System

7.1 Introduction

Many airports worldwide, particularly large airports, are increasingly being faced with requirements relating to approval of plans for expansion of airside and landside infrastructure to accommodate perceived traffic growth efficiently, effectively, and safely. This implies that each such expansion, in order to get approval, must be at least environmentally neutral in terms of increased local noise, emissions of greenhouse gases, airside and landside congestion, land take (use), and safety in terms of traffic incidents/accidents.

In the airport landside area, one aspect of infrastructure expansion in such an environmentally neutral (if not friendlier) way, is related to airport ground access systems. Two European airports are the most illustrative examples of such developments. One is London Heathrow Airport (UK) where approval for building the new Terminal Five was given at least partially due to the agreement between the airport and the environmental regulator on improving rail links between the airport and the city of London. As a consequence, Heathrow Express Link was built [7]. In addition, this has included designing and subsidising new local bus routes mostly serving airport employees. The second case is Zurich Airport (Switzerland), where approval for capacity expansion was given after the Airport's guarantee that the rail access market share would be increased from 34 to 42% during the period covered by the capital investment. This also included developing a local tram line in the vicinity of the airport to serve employees. Consequently, in the given context, two opposing requirements have always been traded-off: increasing the transport capacity of these systems, but not on account of increasing the environmental impacts-burdens. Solutions have implied reducing use of individual cars/taxis and increasing the use of public transport systems in the market segments consisting of air passengers, airport employees, and visitors. In many cases, various rail-based systems (conventional local and national, LR (Light Rail), and/or even HSR (High Speed Rail) have represented a rather strong part of the solution.

M. Janić, *Greening Airports*, Green Energy and Technology,
DOI: 10.1007/978-0-85729-658-0_7,

This chapter describes the environmental feasibility of the LRRT (Light Rail Rapid Transit) system operating as the ground access system of a large airport. The environmental feasibility implies capability of the LRRT system to, under given circumstances, contribute to maintaining or diminishing the total externalities generated by the airport ground access systems in terms of local noise, emissions of greenhouse gases, road congestion, and traffic accidents/incidents on the one hand, and providing additional transport capacities for both air passengers and airport employees on the other.

The main hypothesis to be accepted or rejected is whether a dedicated LRRT system connecting a given airport to its downtown area or CBD (Central Business District) is able to fulfil the above-mentioned requirements and expectations and to what extent, thus contributing to overall greening, i.e. more sustainable development of a given airport. This chapter describes a possible way of dealing with this hypothesis. As will be shown, this includes an analysis of airport cases where rail-based systems, including LRRT, have been successfully used for ground access, specifying the requirements for LRRT operating as the airport ground access system from the aspect of particular parties involved, developing a methodology for assessing the environmental feasibility of LRRT, and application of the methodology to the given airport case according to the "what-if" scenario approach, particularly taking into account the influence of gradual introduction of electric cars. The results, in terms of savings in particular environmental impacts-costs considered as externalities, can be used as inputs on the effects–benefits side for evaluating the overall social-economic feasibility of the LRRT system in the given context.

7.2 Airport Ground Accessibility

7.2.1 Background

Airport ground accessibility is provided by different ground access systems/modes, consisting of private and public transport systems/modes. The former usually include private cars and taxi-cabs, while the latter include different types of bus services (charter, regular, rapid, semi-rapid) as road-based systems, using streets, road links, and highways to transport users to and from the given airport. In addition, the public transport system includes rail-based systems such as conventional and dedicated sub-urban rapid transit rail, and conventional and HSR (High-Speed Rail) connecting a given airport to national and international rail network(s) (see Chap. 4). The former rail systems use rail links connecting the airport and its catchment area or CBD (Central Business District) or the city centre of the main city served by the airport. At large airports, the services of all mentioned ground access systems/modes are provided. At smaller regional airports, services by only rail-or road-based public systems, and/or by taxicab and private car are available. In both cases, the main users are air passengers and airport employees. In addition, users may also include airport visitors and potentially also other commuters.

Convenient interfaces are provided at both ends of the particular airport ground access system. At the airport side, these enable users passage from the given ground airport access system to the airport terminal building(s), and vice versa. In general, these include parking areas for private cars and taxis (with loading and unloading platforms); rail and bus stations (terminals); escalators and moving walkways. In addition, these also include intra-airport transport systems, which usually operate at large airports to provide passengers efficient transfer between distant terminal buildings. These systems include minibuses, standard buses, long moving walkways, and people movers. At the CBD or city centre, interfaces between airport ground access systems and urban transport systems enable efficient and effective passing of users between them in both directions (these include waiting platforms, walkways, conveniently deigned paths for moving baggage, escalators, spaces for short car/taxi stops/parking for picking-up/dropping-off users, etc.).

7.2.2 Accessibility Problems of Large Hub Airports

The above-mentioned cases of two European airports have suggested that efficient and effective ground access systems for large airports should inevitably include public transport and preferably rail-based systems. But has this always been the case? Some research on experience so far indicates that there should also be some other necessary attributes at the given airport in order to make the market share of public transport systems there relatively substantive [21].

As mentioned above, airport size is important. However, despite the fact that a given airport needs to be relatively large in terms of passenger traffic volumes, size alone does not guarantee a substantive market share of public ground access systems.

The distance between a given airport and the core of its gravitational area (CBD or city-centre) may have an opposite effect, i.e. shorter distance discourages use of public transport systems, if car-taxi services are relatively affordable, and vice versa.

The speed of the line-haul vehicle between the airport and city-centre also plays a significant role in choosing public transport system(s), but as in the case of already mentioned attributes, this is not proved to be of critical importance.

Lastly, fewer or even no transfers between particular transit systems along the airport access route are seemingly preferable.

Table 7.1 gives the market share of rail-based systems at selected European airports including travel time, speed ratio as compared to private car, and distance to CBD (city-centre). In addition, information about type of rail service, i.e. dedicated or not, is provided.

Table 7.1 Some characteristics of rail-based ground access modes of selected European airports

Airport	Rail market share (%)	Rail travel time (min)	Ratio car/rail time	Distance (km)	Dedicated service
Zurich	42	20	2	7	No
Oslo	39	19	2.6	48	Yes
Amsterdam	35	17	1.8	15	No
Copenhagen	33	13	1.0	11	No
Munich	31	41.10	1.1	27	No
Vienna	30	16	1.0	19	Yes
Paris CDG	28	35	1.3	24	No
Paris Orly via *People Mover*	14	35	0.7	22	No
London Stansted	29	40	1.7	56	Yes
London Heathrow					
Express	9	15	3.0	24	Yes
Tube	14	45	1.0	24	No
London Gatwick	20	30	2.7	48	No
Geneva	21	10	1.0	4.8	No
Frankfurt	27	12	1.7	10.0	No
Dusseldorf	18	12	1.0	8	No
Brussels	16	14	1.4	11	No
Stockholm	18	20	2.0	40	Yes

Compiled from [14]

It is easy to observe across these airports that there is no strong relationship between the rail market share and factors such as rail travel time, ratio of this time as compared to that of car, and travel distance in case of both dedicated and non-dedicated services. It seems that other airport-related specific factors have a rather crucial influence on the choice of rail-based access systems. Nevertheless, the above-mentioned cases offer insight into the main operational characteristics of rail-based systems at these airports. These characteristics mainly include the line-haul time depending on the travel line-haul travel distance, travel speed, which needs to be comparable to that of private car/taxi, dedication of services, efficient connectivity, and diversity of choice of other urban transport systems in the CBD (city-centre) enabling collection/distribution of users. In general, distance can vary, whereas the line-haul speed should be as high as possible along any distance. Convenient interfaces and schedule coordination between airport rail and other urban transport systems in the CDB provide for the efficient and effective exchange of modes and related distribution/collection of users. All these factors contribute to minimising the door-to-door travel time/cost and to making rail-based systems in the given case a relatively attractive alternative compared to private car/taxi services. In particular, dedicated rail services provide guaranteed and even a superior quality of service compared to private car or taxi in terms of reliability of transit time, but allow users to choose further transport systems to/from home in the CBD (city-centre). In each case, they compete with cars or taxis, both usually using the highway network and its links connecting a given airport and its catchment area.

7.2.3 Preferences for New Airport Ground Access Systems

New airport ground access systems are usually expected to simultaneously fulfil the preferences (criteria) of the particular parties involved, such as prospective users (air passengers, airport employees, and other visitors), system investors and operators, the airport itself, as well as the regional/local and sometimes national authorities.

7.2.3.1 Users

Regardless of which category they fall into (mainly air passengers and airport employees), users of the new airport ground access system usually require local accessibility of the system, handling baggage conveniently, service regularity and reliability in terms of execution and travel time, respectively, a reasonable price as compared to those of alternative systems, safety, and security.

Local accessibility implies a reasonably convenient time in which a given airport can be accessed from an origin in its catchment area (CBD-Central Business District or the main city-centre), or vice versa. This access time includes walking, or taking the car or another urban public transport mode such as the bus, tram, and/or metro. In case of car use, short-term parking spaces for picking-up/dropping-of users should be available near the airport, while in case of using public transport, diversity should exist as regards access system choice, coordinated schedule, a minimum transfer time, and appropriately designed walking paths for efficient moving of baggage (this is important for airport passengers as users of a given combination of systems). In general, reasonable access time via any public and/or private urban transport systems seems to be about 15/20 min. The waiting time for interchange depends on the frequency of services of the given combination of airport access and urban transport systems.

Convenience of handling baggage may include the possibility of using an integrated baggage and ticketing process including advance check-in at the city-centre terminal (station) of the new access system. This generally makes further stages of boarding flights at the airport more efficient and effective.

Regularity implies availability of services over time (hour, day, season, and year). In terms of execution, reliability implies that services are provided without failures, while in terms of travel time, reliability implies carrying out services without significant deviation from the scheduled time due to known reasons.

Service price is expected to be acceptable and comparable to the prices and/or costs of other alterative systems (modes). This can also include a combined price for an integrated ticket/service provided through a combination of different systems.

Safety implies that the system is free from incidents/accidents and consequent damage of property and fatalities caused by known reasons.

Security implies using the system without incidents/accidents caused by other users and/or others.

7.2.3.2 System Investors and Operators

Investors and operators prefer/require the new system to operate profitably. This implies that investors (public/private) receive a return during the period covered by the capital investment. System operator(s) expect profitability from operating the system in terms of covering their operational costs (through revenue, subsidies, and/or both).

7.2.3.3 Airport Operator

The airport operator's preferences may include provision of additional capacity to support further smooth traffic growth, neutral and/or positive direct and indirect contribution to the airport's business, and particularly its contribution to diminishing the overall airport externalities, this time in the landside area.

Contribution to the airport's smooth growth implies that the new system, in addition to providing additional transport capacity, should also enable eventual shortening of airport access time and increase its reliability. In addition, it is expected to prevent overloading of other systems and any consequent deterioration in their quality of service.

Neutrality and/or positive direct and indirect contribution to the airport's business implies that the new system should not adversely affect the airport's overall business in any way. This may relate to the eventual diminishing of airport revenues from car parking, and/or various indirect expenses of visitors, etc.

Contribution to diminishing the overall airport externalities implies that the new system should have lower externalities than existing systems in both absolute and relative terms on the one side and indirectly contribute to diminishing the overall externalities on the other.

7.2.3.4 Regional/Local (and National) Authorities

The regional/local and national authorities/communities will want the new system, if implemented, to not take much new land and cause related landscape deterioration, to be revenue/cost efficient as much as possible, and be environmentally friendly.

The use of (new) land for building and operating the new system is particularly important around large airports located close to their catchment areas, which are already usually heavily populated and infrastructure-built. In such a context, it appears quite difficult to provide additional land (space) for alignment of the new system's infrastructure (rail lines-tracks).

Revenue/cost efficiency implies that the system is able to cover the highest possible share of its operational costs, in which case subsidies from regional/local and sometimes the central government will be lower, if needed at all.

Environmental friendliness implies that the system, as in case of the airport, contributes to diminishing the overall externalities, this time in the scope of the entire regional/community transport sector.

7.3 Light Rail Rapid Transit as an Airport Ground Access System

7.3.1 Background

Light Rail Transit (LRT) or its faster version Light Rail Rapid Transit (LRRT) systems may have different definitions. One of these is that they are electric railway systems with a "light volume" passenger capacity compared to conventional (heavy) rail. Light rail may share or exclusively use the rights-of-way high or low platforms enabling passengers to easily embark/disembark single or multi-car trains. The system is mostly used in urban and sub-urban areas. Currently, there are 24 LRT systems in operation in the US. In 2001, the number of initiated projects (12) was even greater than the number of conventional (heavy) rail projects (7). In Europe, LRT systems are often considered together with urban tramway systems. Some evidence indicates that there are about 170 tram and LRT systems in Europe, encompassing 941 lines of a total length of 8,060 km. Currently, in 21 cities, 154 existing lines are to be extended in total for about 154 km. In addition, 21 new lines of a total length of 455 km are under construction [7].

At many European airports, no LRT systems are currently in operation. As can be seen in Table 7.1, these systems are exclusively based on conventional (heavy) rail technology. In the US, at two airports Baltimore/Washington (Baltimore) and Lambert (St Louis), LRT is in operation as the airport ground access system.

Baltimore/Washington airport is located about 17 km from the centre of Baltimore and about 51 km from the centre of Washington, D.C. The area around the city of Baltimora has a population of 5.6 million and the area around Washington a population of 4.2 million. In 2008, the airport served 20.2 million passengers, of which about 34% were O&D passengers. The airport has a variety of ground access connecting services. One of these is the LRT system connecting the international air terminal to Baltimore centre. During the day, from 8:00 am until 11:00 pm, services are scheduled every 30 min. Each service takes about 23 min to reach the Baltimore CDB or city centre, and vice versa, which, taking into consideration the distance of 17 km, gives an average commercial speed of about 44 km/h. Nevertheless, the market share of the LRT system is well under 1%, while the total share of public transport systems amounts to about 12%.

The main influencing factor is the use of public transport, both bus/van and rail, to access the more distant centre of Washington D.C. In addition, the LRT system does not exclusively serve airport users but also other commuters in the area [21].

Lambert–St. Louis International Airport is located about 21 km from the centre of St. Louis (population about 2.6 million). In 2008, the airport served about 14.5 million passengers, of which about 34% were O&D passengers. The airport is connected by the LRT system to St Louis city-centre (Red Line). There are two LRT stations at the airport. The first station is integrated into the airport's main terminal, enabling users to access the system under the 'common roof', while the second station serves passengers from the East terminal predominantly travelling with Southwest Airlines. LRT services depart and arrive at the airport every 20 min during the day, between 08:00 am and 11:00 pm. Consequently, the average commercial speed of service is about 57 km/h. The market share of public transport in airport access amounts to about 6%, of which 3% is carried by LRT and the rest by bus/van. Some estimates indicate that operational costs of the LRT system amount to about 0.10 €/p-km and total costs to about 0.39 €/p-km. This LRT system is not dedicated to exclusively serving airport users but also commuters in the St. Louis area [21]. As with any other airport access system, the LRRT system can be considered in this case through multidimensional examination of its infrastructural, technological, operational, economic, environmental, and social performance.

7.3.2 The Light Rail Rapid Transit System Performance

7.3.2.1 Infrastructural

As an airport ground access system, LRRT can operate as a dedicated and non-dedicated right-of-way system. In the former case, the system serves exclusively airport users in the wider sense such as air passengers, employees, and airport visitors whose origins and destinations are the airport and the CBD or city-centre of the airport catchment area. In such case, the line has terminals (stations) only at both ends of the line, i.e. at the airport and in the CBD. In the latter case, the system serves, in addition to airport users, also commuters travelling between places along the line, as described in cases of the above-mentioned US airports. In this case, stations, which are easily accessible on foot, by car/taxi, and/or by other urban public transport system, need to be set up along the line.

Typically, a line connects a given airport to its CBD or city-centre area rather than a network of lines. The length of the line is the distance between the start and end terminal (station). The typical width of the corridor for a double track line respecting LRRT vehicles' dynamic envelope amounts to 7.5 m [24]. The typical area of the station's platform along the line amounts from 12 × 50 m (surface) to 20 × 90 m (grade separated). The spacing of stations along the line of a non-dedicated system usually varies between 350 and 800 m [3, 24]. In general, the total area of land taken for the line and stations/platforms can be estimated as:

$$S = 7.5d + ms \tag{7.1}$$

where

d is the length of the line (km);
m is the number of stations/platforms along the line;
s is the area of each station/platform (m^2).

The track gauge is 1,435 mm. The minimum curve radius is 35 m for the yard track and 35 m for the main line. The vertical curve radius is 500 m. The maximum absolute operating grade is 6%.

7.3.2.2 Technical/Technological

The technical/technological performance includes the technical/technological characteristics of LRRT vehicles, which are based on the U2 Siemens Frankfurt (or S70) design as follows [3, 19, 24]:

The vehicle's dimensions: length: 24.2 m; width: 2.78 m; and height: a) 4.0–6.9 m (including pantograph). This vehicle is suitable for passenger embarking/disembarking over a high platform, which makes it convenient for operation as part of the airport access system.

The vehicle's carrying capacity and weight can vary as follows:

Empty vehicle:	32.6 t
Empty vehicle + driver + 65 seated passengers:	37.0 t
Empty vehicle + driver + 161 passengers:	43.6 t
Empty vehicle + driver + 211 passengers:	43.6 t
Empty vehicle + driver + 259 passengers:	50.3 t

Since each vehicle has six axes, the average axial load of the heaviest version is 8.33t/axis. In addition, dedicated space needs to be provided within the carriages for storing passenger baggage.

Power

Each vehicle is driven by two electro motors, each using power of 600–750 V, DC, and with 1,200 rpm (Rotations Per Minute). The average energy intensity amounts to between 1.6 and 5.1 kWh/vehicle-km.

7.3.2.3 Operational

The operational performance of the LRRT system includes the moving characteristics of individual vehicles, characteristics of composing train sets from individual vehicles, and characteristics of running transport services along a given line:

The individual vehicle's moving characteristics are as follows:

Speed: maximum design speed: 80 km/h; maximum operating speed: 70 km/h
Acceleration/deceleration: 1.32 m/s^2
Jerk limit (*changes in acceleration and deceleration*): 1.3 m/s^2
Breaking rates: normal: 1.3 m/s^2; emergency: 2.7 m/s^2.

Composing Train Sets From Individual Vehicles

Usually, two vehicles are coupled in a train set. Considering the above-mentioned dimensions of station platforms and the length of a two-vehicle train set (about 50 m), this appears a convenient train configuration.

Transport Services Along the Line

Depending on the volume and intensity of passenger demand, transport services along the line are usually performed at regular time intervals of one, half, or quarter of an hour, or even at ten minute intervals. The above-mentioned service frequency prevails during the day. During the night, services are much less frequent and depend on the scale of airport operation. In most cases, service punctuality under regular conditions is relatively high, usually over 90%. The eventual time deviations from the schedule at the beginning and the end station are relatively small, in the range of a couple of minutes. Such operations, which are maximally adapted to the prospective demand, seem to be promising for making the system competitive to the other airport ground access systems [24].

7.3.2.4 Economic

The economic performance of the LRRT system operating as an airport ground access system usually includes investments and operator costs and revenues. These appear relevant for evaluating financial feasibility over time, for planning and design, as well as for comparison of the system with other systems either in their planning and design and/or operational phase. Both categories of costs are expressed per unit of system output such as veh-km or p-km. Some examples in Europe and the US may be illustrative. In Europe, the investment costs for three LRT systems in Stockholm (Sweden) amounted to 15–20 million €/track-km.

The average operational cost of the systems amounted to 0.10€/p-km. In the US, the average investment cost for 29 LRT systems amounted to 19.7 million €/track km, while the average operational costs for 15 LRT systems amounted to 0.21€/p-km. In both cases, the period of capital investment is 25 years. Frequently, operating revenues do not cover the costs and subsidies are required, in some cases at the level of about 20–25% [8, 20].

7.3.2.5 Environmental

The environmental performance of LRRT operating as an airport ground access system mainly includes energy consumption, air pollution of greenhouse gases and climate change, noise, congestion, and traffic incidents/accidents, i.e. safety (the latter is sometimes categorised as a social characteristic).

Energy Consumption

As mentioned above, the LRRT system consumes electric energy at an average rate of between 1.6 and 5.1 kWh/vehicle-km. This energy can be obtained from different renewable and non-renewable sources. Usually, a combination of sources is used, differing across particular regions and countries. In any case, using renewable sources is more preferable from this perspective.

Emissions of Greenhouse Gases and Climate Change/Air Pollution

The LRRT system does not directly emit greenhouse gases which would contribute to global warming together with local emissions/air pollution. However, the emissions from production of electricity consumed by the system should be considered. In such a context, they are directly proportional to the rate of the system's electricity consumption and the emission rate from electricity production.

Noise

The LRRT vehicles, similarly as the vehicles operated by other transport modes, generate noise, which in this case depends on the distance of an observer from the vehicle as the source and its speed. Figure 7.1 shows an example [3].

As can be expected, exterior noise is higher than interior noise. In addition, both increase approximately linearly with the train's speed. For comparison, urban buses generate external noise of about 87.5–92.5 dB at a speed of 70 km/h and distance of 5 m [1].

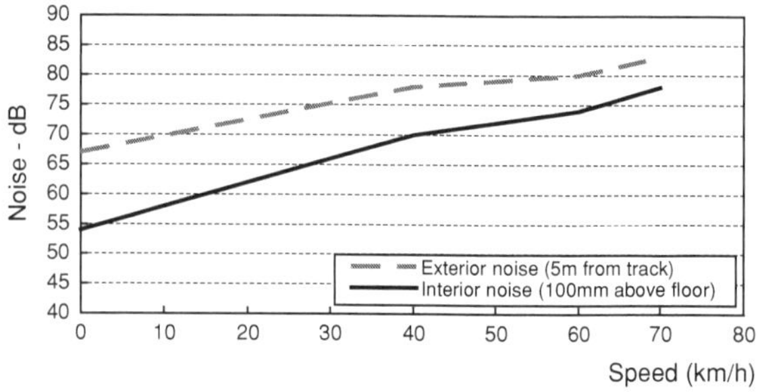

Fig. 7.1 Noise depending on the speed of the LRRT system (Compiled from [3])

Traffic Incidents/Accidents, i.e. safety

The LRRT used as an airport ground access system should be designed and operated free from traffic incidents/accidents due to known reasons, i.e. be absolutely safe.

7.3.2.6 Social

The social performance of the LRRT system operating as an airport ground access system usually implies additional employment and various measurable and non-measurable effects–benefits. Employment stems from the need to plan, design, and construct the system as well as from the need to operate and maintain the system after its implementation. The main measurable effects–benefits expressed in monetary terms are savings in the externalities from other transport systems thanks to diverting, i.e. taking-over, part of their market share. Some non-measurable effects–benefits may include influencing people's behaviour towards more intensive use of public transport systems, providing transportation for users with a wider income range, stimulating further urban and airport development at both ends of the line, and generally diminishing reliance on private car/taxi services (more intensively in the US than in Europe).

7.4 Methodology for Assessing the Environmental Potential of the Light Rail Rapid Transit System at an Airport

7.4.1 Background

In the planning and design stage, each expansion of the airport airside and landside infrastructure capacity is usually the subject of socio-economic evaluation. This also relates to new airport ground access system(s). In general, any such evaluation should include comparison of prospective benefits for particular actors involved, and related costs. In case of a new LRRT system, the prospective benefits include:

- Revenues for the system's operator;
- Users' benefits; and
- Savings in externalities.

The system operator's benefits mainly originate from charging users. In some cases, part of the operator's revenue will be from subsidies. Users' benefits usually include the cost of saved time due to higher line-haul speed of LRRT, convenient frequency, and more efficient connectivity during the system's interchange in CBD (city-centre).

Savings in externalities include the net cost of externalities saved due to implementing the given LRRT system. This implies balancing the reduced externalities from other alternatives and those generated by the new LRRT system.

Costs generally consist of:

- Investment costs;
- Operational costs; and
- External costs (externalities).

Investment costs include costs of building the line (or the network of lines), costs of other facilities (terminals/stations, interfaces) and equipment, and the cost of rolling stock. These costs are usually spread over the period of capital investment, which in the given case amounts to 25–30 years.

Operational costs include the costs of running transport services, i.e. costs of energy, labour, infrastructure and rolling stock maintenance, marketing costs, and administrative costs.

Externalities include costs of noise, air pollution and climate change, traffic incidents/accidents, and land use. Specifically, the cost of land use is usually included in investment costs, while the costs of disrupting landscape, local flora and fauna can be deemed externalities.

The methodology which follows particularly deals with evaluating externalities that could be saved by introducing a dedicated LRRT system as a complement to the exsting ground access systems of a given airport.

7.4.2 Previous Research

The previous academic research has mainly focused on estimating the demand for particular airport ground access systems generated by both air passengers and airport employees. In such contexts, models for estimating the modal split (i.e. market share) of each system have been developed and estimated. Such models have then been used for planning and designing these systems through assessing the volumes of demand for each. These models represent an essential analytical capability for planning and design of these systems, although they have been very varied in terms of functional (analytical) form and variables included. The variables included reflect some attributes of particular ground access systems relevant for users' choice on the one hand and user characteristics relevant for their choice on the other. The first category of variables generally includes accessibility, frequency, regularity and punctuality of services, travel time, and fares as the main attributes for choice of public transport systems (bus, rail). Cost and travel time dominate as the attributes for choice of private systems such as car/taxi. The main user attributes include the value of time and income characteristics.

The development of these models started during the 1970s and 1980s. Research from this period focused on developing and estimating MLN (Multinomial Logit) and NL (Nested Logit) models, the latter expected to overcome limitations of the former. Consequently, the MNL model developed in 1984 for the Washington-Baltimore area included six airport ground access systems and two market segments. The NL model developed for London airports in 1987 included five airport ground access systems and four market segments. In 1998, an NL model for the San Francisco Bay area was developed including nine airport ground access modes serving four market segments. In 1996, an NL model for Boston airport was developed including eleven ground access modes and four market segments. In 1999, a similar model including six airport ground access modes serving four market segments for Portland (Oregon) Airport was developed. In 1999, an NL model was developed in the scope of Integrated Airport Competition Model for Amsterdam Schiphol Airport. Most recently, in 2002, an NL model for London airports including eleven modes and six market segments was developed [22]. These models were estimated (calibrated) by using passenger survey data and mode trip generated data. In general, in addition to the obvious benefits, they possessed some shortcomings. For example, in most cases, they were not tested respecting their predictive, transferability, and flexibility capabilities. These implied the lack of taking into account different circumstances, airports, and regional planning studies. In addition, these models were not tested in terms of flexibility to variations of data and their representativeness for particular airport(s), inclusion/exclusion of the most relevant variables, changes of air travellers' behaviour, and introduction of new systems. At the same time, the same analytical form of the model was mainly used for assessment of the airport ground access mode choice by both air passengers and airport employees. The latter category was discussed exclusively only in few cases [9, 21]. Nevertheless, despite the above-mentioned shortcomings, these models continue to be used by academics, consultants, planners, and designers in airport ground access planning and design.

7.4.3 The Objectives and Assumptions

The objective of this chapter is to elaborate the environmental feasibility of operating a new LRRT system as an airport ground access system. This includes developing a methodology consisting of a series of models and their application to estimating the LRRT system's demand and capacity, revenues and costs, externalities, and the environmental potential as compared to that of the existing ground access systems/modes operating at a given airport under given conditions.

The methodology is based on the following assumptions:

- The medium-to long-term scenarios of volumes of air passengers and the related number of employees as prospective users of the new access system are given. The relative relationships between these two variables are assumed to be stable over the specified period of time.
- Air passenger demand can be segmented into different categories of users of the airport ground access systems. This can be with respect to the users' origins and destinations at the macro scale and the air trip purpose (business, leisure). Airport employees as potential users of the airport ground access system are segmented according to the region(s) of the airport's catchment area where they live. The relative shares of segments of the particular categories of users are assumed to be constant during a given period of time.
- The modal split between particular airport ground access systems is based on airport trip generation data and not on disaggregate passenger survey data;
- Air passengers can use each of the available airport ground access systems regardless of the purpose of their trip (business, leisure);
- Performance of the new LRRT and other existing airport ground access systems/modes, which influences users' choice, are assumed to be relatively constant over a given period of time;
- The externalities from particular airport access systems such as energy consumption, air pollution and climate change, and noise are estimated by taking into account the current-, medium-, and long-term prospective impacts and related costs. This implies that possible impacts from introducing, for example, electric-powered cars and buses are taken into account.

7.4.4 The Basic Structure

7.4.4.1 Model for Estimating Demand for the Light Rail Rapid Transit System

Demand for the new LRRT system operating as an airport ground access system consists mainly of air passengers and airport employees. Estimating the current and future volumes of air passenger demand can be carried out by using the four-stage transport planning model. This includes generation and attraction of passenger flows in particular zones, distribution of these flows among zones of the given region, modal split, and assignment of flows to particular links of the network of each mode. These flows can be further converted into flows of vehicles, which enables the planning and design of related transport infrastructure. In the present case, there are two zones of originating and attracting passenger flows—the airport and CBD or city-centre area. This also implies distribution of all

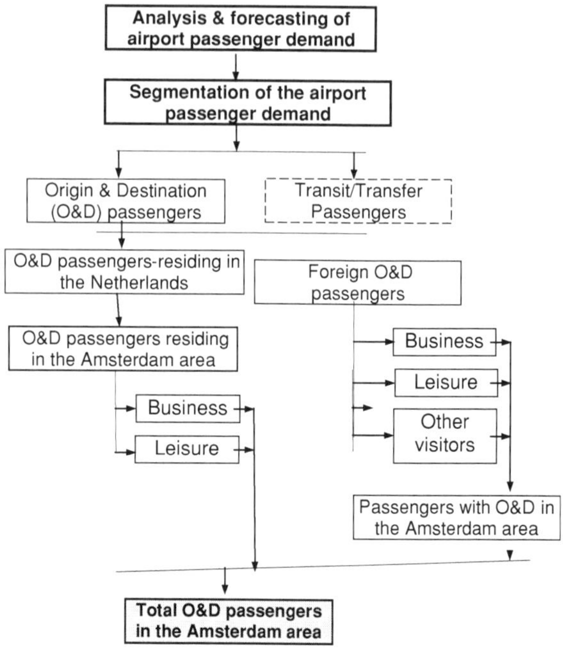

Fig. 7.2 Procedure for estimating passenger demand for ground access systems: The example of Amsterdam Schiphol Airport (The Netherlands)

relevant passenger flows only between these two zones. At large airports with a global impact on the region (country) they serve, estimating these flows for the purpose of planning a new ground access system can be a rather complex task, particularly due to the lack of relevant data. Figure 7.2 shows an example of a possible procedure for estimating the relevant volumes of air passenger demand between Amsterdam Schiphol airport and the Amsterdam greater area (The Netherlands).

As can be seen, their analysis and forecasting start with making a distinction between airport O&D and transit/transfer passengers. The former spreads further along two branches. The first one relates to those air passengers residing in the Netherlands, and particularly in the Amsterdam area. These passengers are distinguished with respect to their trip purpose as business and leisure. The second branch relates to foreign air passengers with different trip purposes who start and finish their trips in the Amsterdam area. In both cases, the specific demand generating/attracting socio-economic attributes are population, GDP (Gross Domestic Product), and tourist attractiveness of the Amsterdam area as compared to the rest of the Netherlands. Their relative relationship and influence at both the regional and country levels can be assumed to be relatively constant despite changes in the absolute values over time. Consequently, the relative proportions of particular categories of air passengers between two zones can also be assumed to be relatively constant despite changes in the volumes of air passenger demand driven by changes of particular demand-driving factors (forces) over time.

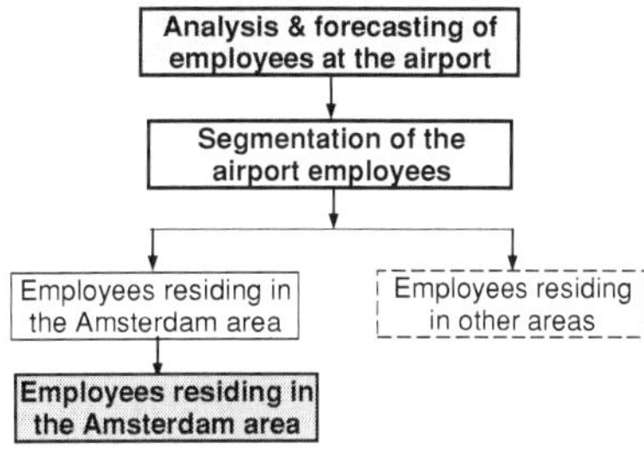

Fig. 7.3 Procedure for estimating demand for ground access systems generated by the airport employees; The example of Amsterdam Schiphol Airport (The Netherlands)

Figure 7.3 shows a procedure for estimating the volumes of demand for the new LRRT in the given example generated by airport employees.

These volumes include airport employees residing in the Amsterdam area. Their proportion is assumed to be relatively constant over the observed future period despite changes in the total number of airport employees driven by changing volumes of air passengers. In addition, this implies that the influencing factors on peoples' choice of place to live will also remain constant during the observed period.

The above-mentioned volumes of both categories of users are split among the available airport ground access systems/modes. The market share of each system/mode can be estimated by using the above-mentioned MNL (Multinomial Logit) or NL (Nested Logit) model [22]. In such contexts, users within each category are assumed to behave similarly while making a choice of which airport ground access system to use. This choice is based on the evaluation of each system's attributes while respecting one's own characteristics and trip purpose. In any case, for particular categories of users and their trip purpose, the disutility function $U_i(T)$, which generally reflects the generalised cost of accessing a given airport via a particular ground access system (i), needs to be estimated. This estimation can be carried out either by using aggregated trip generation data or disaggregate passenger survey data, both for a given period of time (T) [22].

The disutility function $U_i(T)$ may have a different structure adapted to the specificity of particular airport ground access systems/modes. Consequently, the probability of choosing the system/mode (i) at time (T) can be estimated as follows [21]:

$$p_i(T) = \frac{e^{-U_i(T)}}{\sum_{i=1}^{i} e^{-U_i(T)}} \tag{7.2}$$

The number of users choosing the system/mode (i) during the period (T) can be estimated as follows:

$$Q_i(T) = p_i(T) Q(T) \tag{7.3}$$

where

$Q(T)$ is the total number of users of the airport ground access systems during the period (T).

7.4.4.2 Model for Estimating Light Rail Rapid Transit System Capacity

The main planning, design, and operational characteristic of LRRT as a rail-based airport ground access system/mode is the line capacity. This is defined as the maximum number of transport units (or train sets) which can pass through a fixed point on the line (i.e. the "reference location") in a given period of time T (usually one hour) under conditions of constant demand for service [24]. This capacity, defined as the maximum service frequency, can be estimated as follows:

$$f_{\max}(T) = \left[\frac{T}{\max(H_{w\min}; H_{s\min})}\right] \quad (7.4)$$

where

$H_{w\min}$ is the minimum headway between successive trains along particular sections of the line (min);

$H_{s\min}$ is the minimum station headway defined as the inter-arrival time of successive trains at the particular stations along the line (min).

In most cases $H_{w\min} > H_{s\min}$, thus station headway determines line capacity. Consequently, the vehicle line capacity expressed as the maximum number of vehicles which can pass through a given "reference location" during a given period of time can be estimated as follows:

$$C(T) = f_{\max}(T)n \quad (7.5)$$

where:

n is the number of vehicles composing a train set.

The offered capacity of the line C_0, defined as the number of passenger spaces based on expression (7.5), can be estimated as follows:

$$C_0(T) = C(T)N \quad (7.6)$$

where

N is the number of passenger spaces per vehicle.

In addition, the maximum number of train sets for operating along the line during the period T, can be estimated, based on expression (7.5), as follows:

$$n_{ys} = f_{\max}(T)\tau \quad (7.7)$$

where

τ is the turnaround time of the LRRT train along the line including running time, and stop times at start, intermediate, and end stations (min).

7.4.4.3 Model for Estimating Light Rail Rapid Transit System Costs and Revenue

The total costs of operating LRRT as an airport ground access system consist of investment costs and operating costs. They can be estimated as follows for the period of one year:

$$C_T = C_I + C_o = A + 2 \times 365 f c_f \tag{7.8}$$

where

A is the annual annuity paid for the capital investment in both infrastructure and rolling stock (€);

f is the average daily frequency of service in a single direction; and

c is the average cost per frequency (€/departure).

In expression (7.8), the frequency f is set up to satisfy the expected demand given the train's capacity and the average load factor. In addition, the cost c contains the above-mentioned cost components, which depend on the prices of inputs such as material, labour, and energy.

Revenue from operating the system over the period of one year can be estimated as:

$$R = 365 V_p p + s_u \tag{7.9}$$

where

V_p is the daily number of users of the given system/mode (passengers/day);

p is the average price (€/pax); and

s_u is the annual level of subsidising the given system.

In addition, the annual profitability of the given system/mode can be estimated as the difference between revenue (7.9) and costs (7.8).

7.4.4.4 Model for Estimating Externalities of the Light Rail Rapid Transit System

Externalities from a new LRRT system, as well as from other airport ground access systems, generally include noise, congestion, energy consumption, air pollution and climate change, and traffic incidents/accidents. Each of these can be quantified for a given period of time as the product of the intensity of externality per unit of system output, the volume of system output and the perceived cost per unit of a given externality as follows [12]:

$$C_{e/ik} = r_{ik} V_i(T) c_{ek} \tag{7.10}$$

where

r_{ik} is the rate (i.e. intensity) of generation of externality (k) by a given airport access system (i) (quantity per pax-km or s-km);

$V_i(T)$ is the volume of output of a given airport access mode (i) during the period (T) (pax-km or s-km/day, month, year); and

c_{ek} is the cost of externality (k) per unit of quantity (€/unit of externality).

7.4.4.5 Model for Estimating the Environmental Potential of Light Rail Rapid Transit Systems

Depending of the relative value of its dis(utility) function (expression 7.2), a new LRTT system may take over a certain volume of users from other airport ground access systems and consequently cause modal shift. Let MS_j be the market share of the system (j) before introducing LRRT as a new system (i) and $MS^*_{j/i}$after its full implementation. Let $\Delta MS_{j/i}$ be defined as:

$$\Delta MS_{j/i} = MS_j - M\,S^*_{j/i} \tag{7.11}$$

If $\Delta MS_{j/i} > 0$ the system (j) has lost part of its market share, otherwise it has gained market share or maintained its market share unchanged. The number of users shifted from the system (j) because of influence of the system (i) can be estimated as follows:

$$\Delta Q^*_{j/i} = Q_i \Delta MS^*_{j/i} \tag{7.12}$$

Consequently, the market share gained by the system/mode (i) from all other systems/modes can be estimated as:

$$\Delta Q^*_i = \sum_{\substack{j=1 \\ j \neq i}}^{J} \Delta Q^*_{j/i} \tag{7.13}$$

where

J is the number of airport ground access systems/modes.

The total externalities of the system/mode (k) for a given period of time can be estimated as follows:

$$E_i = \Delta Q_i^* d_i \sum_{k=1}^{K} r_{ik} c_k \tag{7.14}$$

where
K is the number of types of externalities.

Other symbols are analogous to those in the previous expressions.

7.5 Application of the Proposed Methodology

7.5.1 The Case of a Large Airport: Amsterdam Schiphol (The Netherlands)

The LRRT has been planned and designed as a new additional ground access system to connect Amsterdam Schiphol Airport to the Amsterdam area. In general, the new system is expected to increase the capacity of the airport ground access systems, but in a more environmentally friendly way as compared to alternative systems/modes. One of these alternatives includes widening roads to handle the expected increased congestion, which implicitly suggests the continuous extensive use of private cars [16]. Another alternative includes increasing the capacity of existing train and bus services, although the former already operates at the level of line capacity saturation. Thus, the new LRRT system is not intended to improve the spatial connectivity of currently unsatisfactorily-connected parts of the Amsterdam area, but mainly to improve and maintain the required level of connectivity of the already well-connected parts of the area. This also implies that the system should serve sufficient volumes of users and be accessible to the majority of them in the most convenient way. In addition, the total annual volumes of air passengers at Amsterdam Schiphol airport are expected to grow at an average annual rate of about 2.5%, which will result in an increase from the present 47 million passengers to about 80–85 million a year by 2025–2030. The new LRRT system is expected to support the above-mentioned growth [16]. In addition, this line can be part of the LRRT network which is planned to be built to improve the overall accessibility of the Randstad area [13].

7.5.1.1 The Concept of the Light Rail Rapid Transit System

Line Alignment and Design

As shown in Fig. 7.4, a simplified scheme of the right-of-way double-track line will be aligned almost parallel to the existing conventional rail line and/or partially along motorways A10 and A4. It will begin at or near Amsterdam

Fig. 7.4 A simplified scheme of the possible alignment of the new LRRT line between the Amsterdam area and Schiphol Airport (The Netherlands)

Central Railway Station and end in front of or below the main airport passenger terminal.

The line's originating station can be integrated into the existing central rail station, but with separate platforms and tracks, i.e. not shared with those of the conventional railway system. Since trip duration and purpose have shown to strongly influence the quantity of baggage carried by air passengers and consequently their choice of airport ground access system, its design and organisation of passengers' movement, including baggage handling, need to be adequately specified. This implies that the LRRT part of the station should be conveniently connected to the part of station used by the conventional railway system in order to enable efficient and effective passage between the two systems. If the LRRT station is not integrated into the central rail station, convenient and efficient pathways between them need to be provided. In both cases, the LRRT station should have efficient exit/entry and pathway to/from the neighbouring urban public transport systems (taxi, tram, bus, and car parking areas). In addition, easy (automatic) ticketing for LRRT services and off-site advance check-in services will be provided. This requires additional space for installing check-in counters and related equipment such as an automatic baggage conveyer system. The check-in services would be provided from 24 h to about 1.5 h before any given flight. Baggage would be transported to the airport by LRRT train sets equipped with dedicated baggage compartments. Experience has shown that such an off-site check-in system could also be economically feasible if operated by the railways or the airport [21]. Both the airport and dominant airline(s) such as SkyTeam and their partners could also organise such a service.

At the airport, the LRRT station could be either an underground or surface construction. In the former case, it would consist of platforms and tracks, again not shared with those of the conventional railway system. However, passage to/from the main airport passenger terminal would be possible by escalators and lifts, as is the case with the existing conventional railway system. Signage to/from the terminal arrival/departure hall should be clear and enable efficient movement of users. In the case of surface construction, the LRRT station, if not integrated into the main terminal, should preferably be located within walking distance. In such a case, connection to the main terminal would be provided by moving walkways in order to make movement as efficient as possible. Such moving walkways would be closed-off in order to protect users from bad weather. If the LRRT station is located at a greater distance than the walking distance from the main terminal, a people mover system could provide the required efficiency of connectivity. However, this would need additional interchange and consequently potentially diminish the attractiveness of the LRRT system. In any case, airport employees using the system should be ensured convenient passage to/from their working places regardless of the system they use (on foot, by bike, local van/bus, etc.). At both ends of the line, eventual extension should be possible. This should also include intra-airport extension if an additional dislocated passenger terminal is to be built in the future [16].

The line length will be 15 km. Consequently, construction of the LRRT as a right-of-way system with the above-mentioned characteristics will take at least 11.5 ha of new land (see expression 7.1).

Operation

The following main operational characteristics of the new LRRT system could make it potentially attractive to both air passengers and airport employees: convenience of access, waiting time for service, travel time, punctuality, internal comfort, and price. Convenience of accessibility in Amsterdam city-centre implies reaching the system's station from local origin by any means in about 15–20 min, and vice versa.

Waiting for the LRRT service depends on its service frequency, which would be one every 10 to 15 min during the entire day. This implies an average waiting time for departure of about 5–7.5 min. The capacity of an LRRT train set would be between 160 and 210 sitting and standing passengers [19]. Allowing air passengers to check-in in advance could contribute to a more comfortable trip and more effective transport to the departure gate/flight at the airport.

The travel time would be about 15 min, which implies an average travel speed of 60 km/h for this dedicated non-stop service. Under regular operating conditions, punctuality of services would be maintained at the highest level, at about 98–99%. The maximum allowed deviations from the schedule would be in the order of a couple of minutes. Stops at the originating and destination stations would be for about 2.5 min, enabling an average turnaround time of a train set of about 35–40 min. Consequently, 3–4 LRRT train sets would simultaneously operate along the line. The price of service in a single direction would be comparable to that of conventional rail, i.e. 3.5€/passenger.

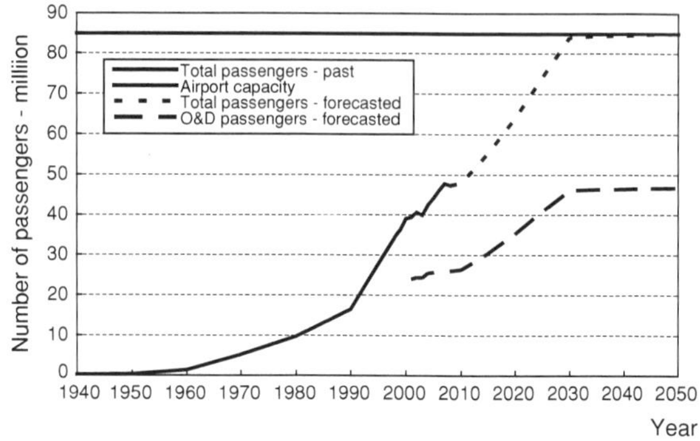

Fig. 7.5 The long-term development of air passenger traffic at Amsterdam Schiphol Airport (1940–2050) (Compiled from [16])

7.5.1.2 Analysis of and Forecasting Demand for the Light Rail Rapid Transit System

Total Airport Passenger Demand

The above-mentioned operations of the new LRRT system are only possible if there are sufficient volumes of demand. These are estimated for both air passengers and airport employees using the procedures in Figs. 7.2 and 7.3, respectively. The results are shown in Figs. 7.5 and 7.6, respectively.

Figure 7.5 shows the past and future development of airport passenger demand. As can be seen, the total annual volumes grew to about 47 million passengers in 2008. About 30 million passengers were carried by members of the SkyTeam alliance (KLM), while 5.1 million passengers were carried by LCCs (Low Cost Carriers). In 2003/2004, the approximate structure of passengers regarding the trip purpose was: business 35%, leisure 42%, studies 3%, visiting friends and relatives 19%, and other 1% [15]. The total volumes of future air passenger demand are assumed to continue to grow, although at lower annual rates, and reach the annual level of about 80–85 million by 2030, and remain at that level (as mentioned above, the long-term growth rate of airport passenger demand is assumed to be 2.5–3.5%/year; after approaching saturation of planned capacity, these rates are assumed to be 0.25–1%/year) [16]. About 40–45% of these total volumes are transit/transfer, while the rest are O&D passengers.

If this proportion remains relatively constant, the annual number of O&D passengers will increase to about 45 million by 2030 and remain at that level in the future. These volumes will be handled by the passenger terminal's existing capacity of about 60–65 million/year. This capacity appears sufficient to accommodate air passenger demand until about 2020.

The above-mentioned airport growth is based on the assumption that the dominant airline alliance SkyTeam and its partner KLM will continue to carry air passengers as they did in the past. This role can change and consequently compromise the predicted growth due to several reasons. The most important/certain ones are: reducing the number of short-haul flights replaced with forthcoming HSR

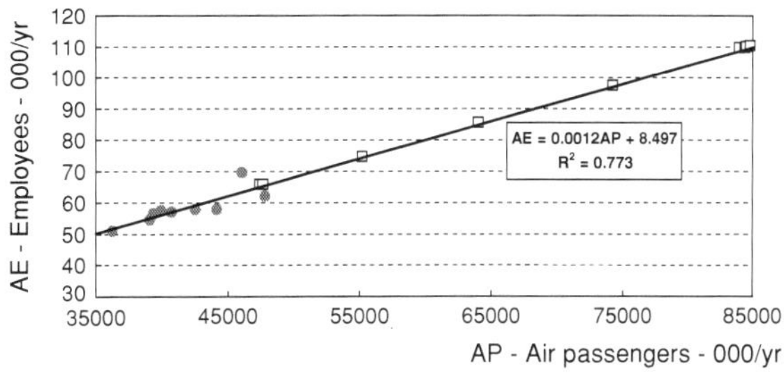

Fig. 7.6 The long-term relationship between air passenger traffic and daily number of employees at Amsterdam Schiphol Airport (1998–2050) (Compiled from [15])

services in 2010, and the changing operational pattern of the SkyTeam alliance by increasing flight concentration at its primary hub—Paris Charles de Gaulle Airport—at the expense of the secondary hub—Amsterdam Schiphol Airport.

The Number of Airport Employees

Based on the above-mentioned analysis and forecasting of air passenger volumes, the corresponding daily number of airport employees for the period 1998–2030 is estimated and shown in Fig. 7.6. As can be seen, their number linearly increased with the annual number of air passengers accommodated during the past decade (1999–2008). The number of airport employees is assumed to continue to linearly increase in the future at an average rate of 1,200 employees per 1 million accommodated air passengers. Consequently, if airport labour productivity remains at the present levels, the number of daily employees will reach about 110 thousand when the annual volumes of air passengers grow to about 80–85 million by 2025–2030.

The Potential Demand for Airport Ground Access Systems

The above-mentioned air passenger volumes and the related number of airport employees are further split into particular categories in order to obtain the total demand with O&D in the Amsterdam area (see Figs. 7.1 and 7.2). At the first level, the air passengers are segmented into residents and non-residents of The Netherlands. The former category assumes a share of about 35%. The proportion of these passengers residing in the Amsterdam area is estimated, regardless of trip purpose, according to the share of GDP of the Amsterdam area in the national GDP (about 12%) [18]. (In this case, GDP is considered as the main driving force for air transport demand regardless of trip purpose). The latter category of air passengers has a share amounting to the remaining 65% of the total. The same GDP-based criterion is used for estimating the proportion of foreign business passengers travelling to/from the area. The proportion of foreign tourists arriving by air to the Amsterdam area is estimated directly on the basis of relevant data [2], while the proportion of foreign passengers visiting friends and relatives is

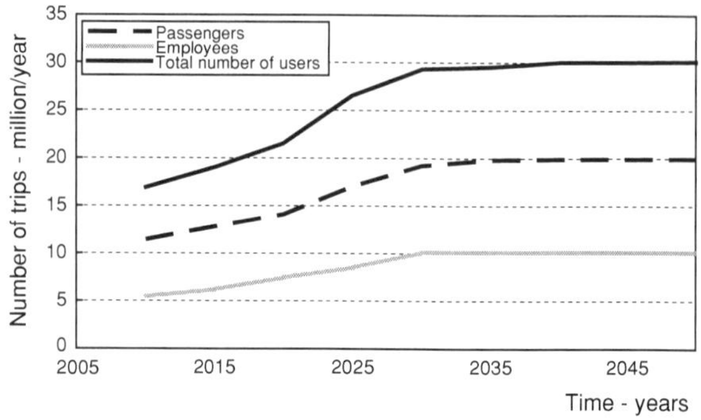

Fig. 7.7 Development of user demand for planning and design of the new LRRT system at Amsterdam Schiphol airport (The Netherlands)

estimated in accordance with the share of the area's population in the total population of the country.

The proportion of airport employees residing in the Amsterdam area is estimated directly on the basis of relevant data (about 18–19%) [15].

Summing up the above-mentioned proportions produces an estimation of the total number of air O&D passengers and airport employees in the Amsterdam area. Assuming that these proportions will remain relatively constant in the future, the annual number of O&D passengers and airport employees from the area is estimated based on the forecasted total airport passenger volumes and shown in Fig. 7.7. In particular, airport employees are assumed to commute to/from the airport every day for about 230 working days of each year of the observed period (non-working days during the week and holidays are excluded).

As can be seen, the number of prospective users of the airport ground access systems/modes including the new LRRT system is expected to grow according to and in proportion to the growth of total air passenger volumes. Consequently, the prospective annual number of trips by users of the airport ground access systems to/from the Amsterdam area could increase by 2030 to about 30 million, of which about 20 million will be air passengers and the rest airport employees.

Prospective Modal Split: Demand for the New LRRT System

In general, the "what-if" approach is applied to estimate the market shares of particular airport ground access systems. Trip generation data for each existing and new system is used to estimate their disutility functions representing the perceived generalised travel cost for particular categories of users—air passengers and airport employees. The method of trip generation is applied assuming that at this stage, planners are able to make sufficiently precise estimates, particularly for the new non-existing system (LRRT). Due to maintaining the internal consistency, the same method is applied to existing systems. In both cases, generalised costs consist of out-off-pocket travel costs and the cost of (door-to-door) travel time. The

Table 7.2 Characterisation of ground access systems at Amsterdam Schiphol Airport

Mode	Access time (min)	Frequency (dep/h)	Distance (km)	In-vehicle time (min)	Average cost (€)
Car	5	–	15	25–50	0.46/km[2]
Taxi	5–10	–	15	25–30	40–45
Bus[a]	10	6	26	30	6.75[b]
Local Rail	20	4–6	15	17.5	3.80 6.40
LRRT	20	1–6	15	15	3.5

[a] Lines: 370, 198, 300, 10, 199,197, 97, 358, 192,195, 61, 188, Airport Shuttle; Source: SG 2008 [15]
[b] Based on the price of fuel of 1.35€/l (middle class car)

distinction between particular market segments in terms of the trip purpose is addressed through specifying the average value of time for the assumed mixture of business and leisure air travellers using a given system. Airport employees are considered a rather homogeneous market segment with the same average value of time. The characteristics of particular systems relevant for choices of both categories of users are given in Table 7.2.

In particular, the access time for car and taxi implies time of walking to/from the vehicle, and/or waiting for its arrival, respectively. The access time of particular public transport systems includes time from the user's place of residence to the nearest station, and vice versa, and the time for passing platforms at the railway stations. In this case, car/urban taxi and/or urban public transport systems-bus, tram, and/or underground-can be used. Alternatively, if convenient, biking and/or walking may be used. The departure frequency reflects the current level of service of existing airport ground access systems. LRRT frequency is the variable parameter while estimating the corresponding disutility functions. The travel distance is assumed to be approximately the same for all systems. In-vehicle time for road-based systems is assumed to take into account congestion. The current fares of public transport systems and the costs of car use are also specified. The average value of time of an air passenger regardless of the trip purpose is adopted to be 0.847€/min while using car and/or taxi, and 0.707€/min while using bus, local rail, or the LRRT system. This reflects the fact that lower income travellers and tourists may be more inclined to use public transport systems [6, 21]. In particular, users of public transport systems are assumed to equally value their time while waiting for a departure and while travelling. The users of private cars are assumed to be picked-up and dropped-off, therefore the parking costs at both ends of the route are negligible. In addition, the difference in the costs of foreign users using hired cars and domestic users as car owners is not particularly considered.

With slight modification, the characteristics of particular airport access systems in Table 7.2 are also used for estimating the disutility functions for airport employees. The modifications mainly include slightly reduced fares of public transport services due to various discounts, reduced in-vehicle time of car users, and the unified value of time across particular systems of 0.291€/min based on employees' average salaries [15].

Table 7.3 Disutility functions for particular categories of users of airport ground access systems in the given example

Disutility function	Air passengers	Airport employees
U_{car}	11.775	10.128
U_{taxi}	20.701	45.460
U_{bus}	14.729	11.984
$U_{local\ rail}$	12.361	11.060
U_{LRRT}	12.252[a]	11.220[a]

[a] Based on the frequency of 4 dep/h

Particular disutility functions while using particular airport ground access systems in the given case are estimated by the modified SERAS model and shown in Table 7.3 [10].

When the MNL model in expression 7.2 is applied, the market share of taxi in both cases in Table 7.3 appears negligible. However, in reality this is not the case since taxi services have an average market share of about 10% [15]. Consequently, the MNL model in expression (7.2) is applied to other systems excluding taxi. In case of the LRRT system, the hourly service frequency is varied and the corresponding disutility functions estimated. The results are shown in Figs. 7.8 and 7.9. Specifically, Fig. 7.8 shows dependency of the market shares of particular airport ground access systems on the frequency of service of the new LRRT system for air passengers.

As can be seen, without the LRRT system, private car assumes the highest market share followed by local rail and bus. After modest introduction of LRRT services (1–2 departures/h/direction), the market shares of existing modes do not change. However, after increasing the LRRT system's frequency above 2 departures/h/direction, the market share of the LRRT system starts to increase and market shares of the other systems decrease, both at a rather substantive rate. In general, the absolute impact is the greatest on those systems with already high market shares (car, local rail), and vice versa.

Figure 7.9 shows the market shares of particular airport ground access systems depending on the frequency of service of the new LRRT system for airport employees.

As can be seen, again, without the LRRT system, private car assumes the greatest market share, followed by local rail and bus. With introduction of the LRRT system at a substantive frequency, LRRT gains market share at the expense of the other systems, as was the case in air passengers.

Figures 7.10 and 7.11 synthesise the losses and gains of the relative market shares of particular airport ground access systems depending on the frequency of the new LRRT system for both air passengers and airport employees, respectively.

Figure 7.10 shows that in the case of air passengers, the LRRT relative market share gain can exceed 20%. At the same time, private car can lose an equivalent proportion of its relative market share, followed by local rail (about 10% loss) and bus (up to 5% loss).

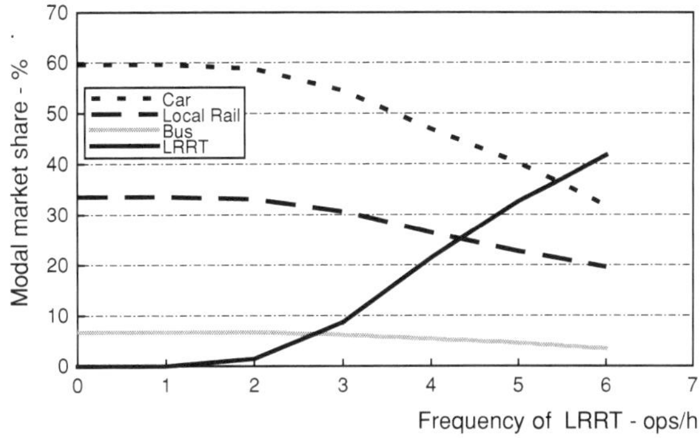

Fig. 7.8 Market shares of particular airport ground access systems for air passengers depending on the LRRT system frequency in the given example

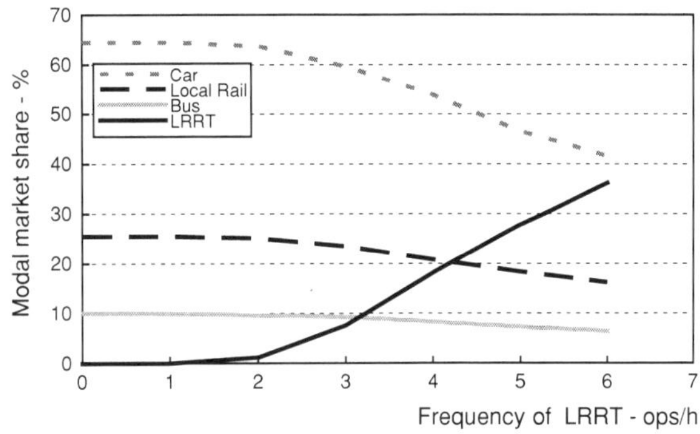

Fig. 7.9 Market shares of particular airport ground access systems for airport employees depending on the LRRT system frequency in the given example

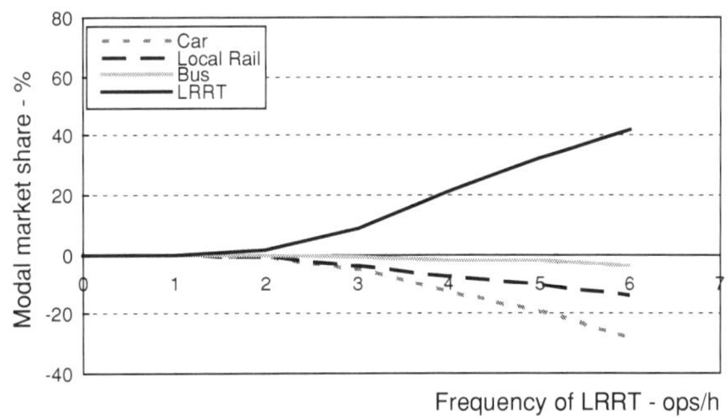

Fig. 7.10 Change in the share of particular ground access system in the air passenger market depending on the service, frequency of the new LRRT system in the given example

Figure 7.11 shows the similar impact of the LRRT system on the relative market shares of other ground access systems for airport employees. However, its gains are slightly less than those in the case of air passengers (18 vs. 21.4% for the service frequency of 4 departures/h/direction). Again, private car records the greatest loss of market share, followed by local rail.

Consequently, the new LRRT system with operational and economic performance comparable to existing airport ground access systems seems to be able to cause a rather substantive modal shift in both market segments—air passengers

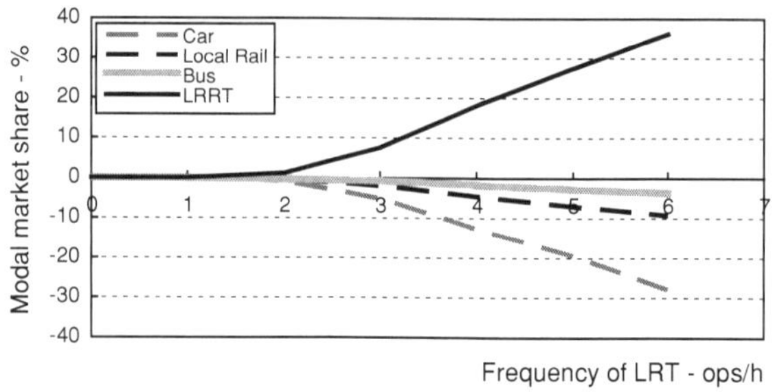

Fig. 7.11 Change in the share of particular ground access system in the market of employees depending on the service frequency of the new LRRT system in the given example

and airport employees. In general, it may affect systems with presently higher market shares such as car and local rail in the given case. Thus, it contributes to the increased use of public transport access systems and particularly those which are rail-based.

7.5.1.3 Matching LRRT System Capacity to Demand

As shown in Figs. 7.10 and 7.11, the LRRT system operating at the average frequency of 4 departures/h/direction can take about 21.4% of air passenger and 18.15% of airport employee volumes. In this case, the LRRT train set can consist of cars with a capacity of 160 or 210 seats, thus offering a seating capacity of between 1,280 and 1,680 seats/h, respectively, in both directions. This capacity is offered during 20 h of daily operations in each year of the observed future period. By comparing the expected volumes based on the above-mentioned shares of the LRRT system in the total corresponding volumes in Fig. 7.7 and the offered capacity, the average load factor is obtained and shown in Fig. 7.12. As can be seen, the average load factor gradually increases with the annual volumes of air passengers and airport employees, while the transport capacity remains constant during the observed period. In addition, if cars with a lower seating capacity are

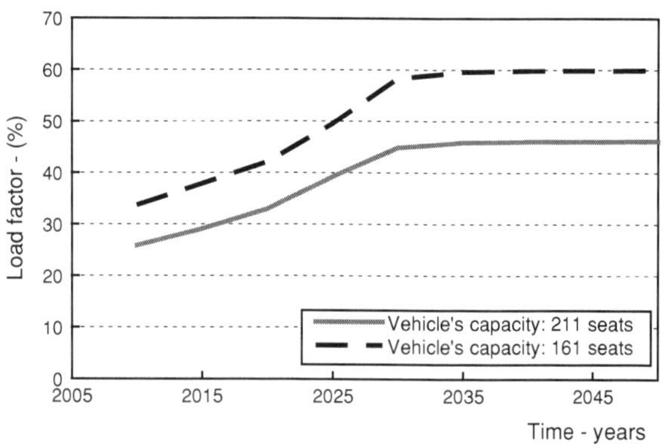

Fig. 7.12 The average load factor of the LRRT system vs. the vehicle's size and time in the given example

used, the load factor will be higher in proportion to the difference in this capacity, and vice versa.

7.5.1.4 Externalities of the Light Rail Rapid Transit and Other Airport Ground Access Systems

The externalities of particular airport ground access systems are estimated using the existing and prospective data on the quantity and per unit cost of particular impacts.

Noise generated by particular airport ground access systems including the LRRT system is discussed in Sect. 7.3.2.5 (see Fig. 7.1).

Congestion is considered only for the road-based-car/taxi and public bus system. The average level of congestion affecting car, taxi, and bus users during almost the entire day is adopted to be 0.70 [11]. This is mainly because the major motorways A4 and A10 are and will continue to be shared by airport and other users. Under such circumstances, an LRRT system with the capacity of a train set of 160 seats with 4 departures/h/direction can replace 320 cars/taxis every hour, thus reducing the average delay of each car for about 1.25 min and of the car user for about 0.625 min (the average car occupancy rate is adopted to be 0.5). The rail and the LRRT system are considered to be free of congestion.

Emissions of greenhouse gases from cars/taxis and public buses are estimated using the corresponding rates of fuel consumption, energy content, and emission factors of CO_2 equivalents. For the given average structure of cars and buses in terms of using particular types of fuel (gasoline, diesel, and natural gas), the average emission rates have been estimated as follows: car—43.5 gCO_2/s-km and bus—47.2 gCO_2/s-km [5, 17, 18, 23]. For the LRRT system, the average emission rate of CO_2 equivalents is estimated using the average rate of energy consumption of the LRRT vehicles (mentioned above) and the emission factor of CO_2 equivalents from electricity production in the Netherlands (this factor amounts to 519 gCO_2/kWh) [5]. Consequently, for an LRRT train set with a capacity of 160 seats consuming electric energy at an average rate of 3.4 kWh/vehicle-km, the average emission rate of CO_2 equivalents is estimated to be 10.91 gCO_2/seat-km.

Traffic Accidents

In the Netherlands, the rate of road traffic incident/accidents and related fatalities has varied, but the generally decreased over time. The national average rate of 47 fatalities per billion p-km in 2008 is applied to car/taxi and bus system [18]. The LRRT system is assumed free of incidents/accidents and related fatalities.

The unit costs of the above-mentioned impacts are adopted to be as follows [4]. For road-based access systems: noise—0.07€ct/p-km; congestion—84.7€ct/min for air passengers and 0.291€ct/min for airport employees (these costs are based on the value of their time); emissions of greenhouse gases and climate change—

0.32€ct/p-km; and traffic incidents/accidents—0.97€ct/p-km. In contrast, for local rail and LRRT systems, the unit costs of impact are assumed to be: noise—0.14€ct/p-km; congestion—0.0€ct/p-km; emissions of greenhouse gases and climate change—0.05€ct/p-km; and traffic incidents/accidents 0.0 ct/p-km.

The above-mentioned estimates of externalities and their relative relationships are assumed constant over the observed future period.

7.5.2 The Prospective Environmental Potential of the New Light Rail Rapid Transit System

The prospective environmental potential of the innovative LRRT system is expressed in terms of savings in externalities due to it taking over the passenger demand from other airport ground access systems. In general, these savings are calculated as the difference between externalities generated by the LRRT system and those saved from other systems due to market share loss and consequent reduction in the scale of their operations. These savings are shown in Figs. 7.13, 7.14, and 7.15.

Figure 7.13 shows the annual savings in noise and emissions of greenhouse gases. In particular, the latter savings are estimated for conditions of no change in the power source of existing petrol/gas powered individual cars and buses. As can be seen, both types of savings amount to tens (noise) and hundreds (emissions/climate change) of million €. In addition, they both increase over time approximately in proportion to the increased volumes of users of airport ground access systems depending on the growth of air passenger and airport employee volumes. In addition, savings in the cost of emissions/climate change are several times greater than those of noise. Nevertheless, in both cases, through the gradual future introduction of quieter and less air polluting electric cars and buses, both savings will likely decrease.

Figure 7.14 shows the prospective influence of the gradual introduction of electric cars on savings in the costs of emissions of greenhouse gases , which can be achieved by introducing the innovative LRRT system.

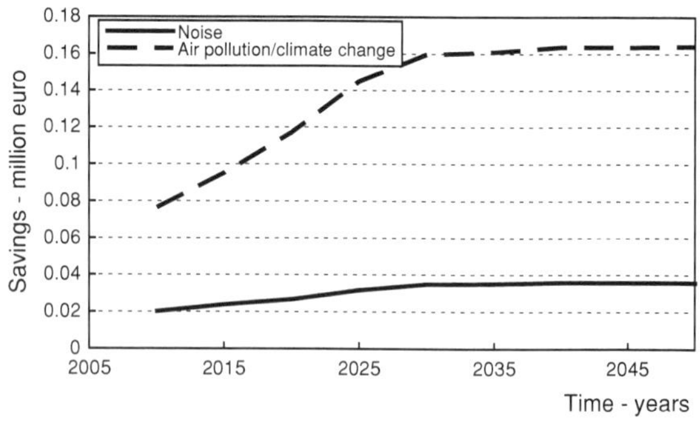

Fig. 7.13 Savings in the cost of noise and emissions of greenhouse gases over time due to the innovative LRRT system in the given example

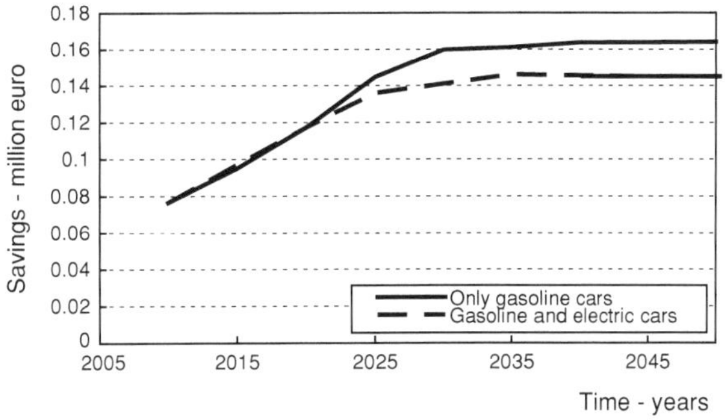

Fig. 7.14 Savings in the cost of emissions of greenhouse gases over time by the innovative LRRT system when electric cars are introduced in the given example

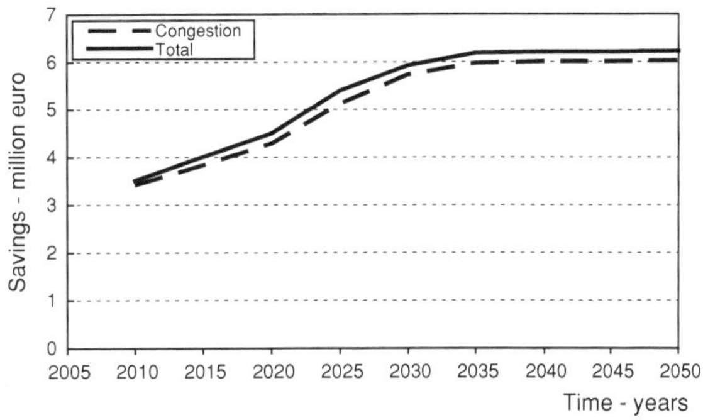

Fig. 7.15 Savings in the cost of congestion and in the total externalities by the innovative LRRT system over time in the given example

As can be seen, a gradual introduction of electric cars at the rate of about 20% for every ten years starting in 2015 would proportionally reduce the expected savings that could be achieved due to the LRRT system. If all cars were electric, the maximum savings in the costs of air pollution/climate change of the LRRT system would reduce by about 23% during the 2055–2060 period. Nevertheless, the overall cost savings achieved by LRRT will remain relatively high. In addition, if the emission rate from electricity production is decreased, these cost savings will decrease proportionally in absolute terms, while their relative difference will still remain noticeable.

Savings in the costs of traffic incidents/accidents are negligibly lower than those of other externalities mainly thanks to the above-mentioned rather low (or zero) rates of all airport access systems.

Figure 7.15 shows the annual savings in road congestion and in the total mentioned externalities over the observed period. Marginal savings from reduced congestion are calculated for those users (air passengers and airport employees) who will continue to use car/taxi and the bus system to access the airport.

As can be seen, the annual savings in cost of congestion range between 3.5 and 6.0 Million € during the observed period. As other externalities, they increase over time due to increased numbers of less affected users of private cars/taxis and public bus systems, mainly thanks to increased volumes of air passengers and airport employees. The total annual savings follow a similar trend. As can be seen, they are only slightly higher than the cost of congestion. This indicates that savings in congestion costs dominate the total cost savings of all impacts. Savings in the costs of emissions/climate change do not particularly influence the total cost savings by the LRRT system also due to the gradual introduction of electric cars.

7.6 Concluding Remarks

This chapter has described the environmental and social potential of an LRRT system operating as an alternative ground access system of a given airport. A methodology consisting of models for estimating demand, capacity, and savings in the costs of particular environmental impacts such as noise, emissions/climate change, traffic incidents/accidents, and congestion has been developed and applied to Amsterdam Schiphol Airport (The Netherlands) according to the "what-if" scenario approach.

The results expressed by savings in the costs of particular above-mentioned impacts have indicated that the new LRRT system may, under given conditions, substantively contribute to greening, i.e. more sustainable long-term development of the given airport as well as its wider catchment area (Randstad, the Netherlands). This system, in addition to efficiency and quality of service offered to prospective users—air passengers and airport employees-seems to be environmentally and socially beneficial mainly due to its contribution to relieving road congestion and related costs and much less due to relieving other impacts and their costs-noise, emissions/climate change, and traffic incidents/accidents. Gradual introduction of hybrid-electric cars and buses will certainly diminish the cost savings in emissions/climate change and eventually noise. However, the savings in costs of road congestion will remain high and dominant, which is the new LRRT system's main contribution. Last but not the least, the above-mentioned savings can be counted as benefits in the overall social-economic evaluation of the LRRT as an additional airport ground access system/mode.

References

1. AUMA (2007) Towards a light rail transit program: analysis and recommendations. Alberta Urban Municipalities Association, Alberta, California, USA
2. CA (2008) Fact sheet: Amsterdam remains a popular tourist destination. City of amsterdam research and statistics, Economic development department. Amsterdam, The Netherlands, p 6
3. CE (2008) Vehicle technology review: city of edmonton. LRTP Projects. http://www.edmonton.ca
4. CE Delft (2008) Handbook on estimation of external costs in the transport sector. Produced within the study internalization measures and policies foe all external cost of transport (IMPACT), CE Delft-Solutions for Environment Economy and Technology, Delft, The Netherlands
5. EC (2009) EU Energy and Transport in Figures. Statistical pocketbook, European Commission, Luxembourg

6. EEC (2007) Standard inputs for EUROCONTROL cost benefit analysis. EUROCONTROL, Brussels, Belgium
7. ERRAC (2004) Light rail and metro systems in Europe: current market perspectives and research implications. The European rail research advisory council, Brussels, Belgium
8. Garrett TA (2004) Light-Rail transit in America: policy issues and prospects for economic development. Federal Reserve Bank, St Louis, USA
9. Gosling DG (2006) Predictive Reliability of Airport Ground Access Mode Choice Models. 85th Transportation research board annual meeting. Session 584, Washington DC, USA
10. HGL (2002) SERAS surface access modeling. Halcrow group Ltd, Report prepared for the department of transport, Local government and the regions, South East and East of England regional air services study. London, England, UK
11. Janic M (2001) Analysis and modeling of air transport system: capacity, quality of services and economics. Gordon & Breach Science Publishers, Amsterdam, The Netherlands
12. Janic M (2009) Light rail system for more sustainable accessibility of large airports. RANSUMO (Transition to Sustainable Mobility) Research Programme, Amsterdam, The Netherlands
13. Kouwehoven M (2008) The Role of Accessibility in Passenger Choice of Airports. Discussion paper No. 2008–14. Joint transport research centre, International transport forum, Significance, The Hague, The Netherlands
14. LFA (2000) Improving Public Transportation Access to Large Airports. Leigh Fisher Associates, MA. Coogan, and Market Sense, TCRP Report 62: TRB, National Research Council, Washington DC, USA
15. SG (2003/2008) Annual reports. Schiphol Group, Schiphol, Amsterdam, The Netherlands
16. SG (2007) Een wereldwijd netwerk voor een concurrerende randstad. Schiphol Group, Schiphol, Amsterdam, The Netherlands
17. SN (2004) Nederlandse lijst van energiedragers en standard CO_2 emissiefactoren. Senter Novem, Utrecht, The Netherlands
18. SN (2008) Statistical yearbook 2008. Statistics Netherlands, Voorburg, The Netherlands
19. STSI (2008) Light rail vehicles. Siemens transportation systems Inc., Rolling stock division, Sacramento, California, USA
20. Tegner G (2004) Comparison of costs between bus, PRT, LRT and Metro/Rail in EC project European demonstration of innovative city transport (EDICT). European commission 5th framework program, DG VII, Brussels, Belgium
21. TRB (2008) Ground access to major airports by public transportation. ACRP Report 4, Transportation research board, Washington DC, USA
22. TRB (2008a) Airport ground access system mode choice models: a synthesis of airport practice. ACRP synthesis 5, Transportation research board, Washington DC, USA
23. VTT (2004) Transit bus emissions study: comparison of emissions from diesel and natural gas buses. Research report. Technical research center of Finland, Tampere, Finland
24. Vuchic RV (2004) Urban transit: operations, planning and economics. Willey, New York

Chapter 8
Conclusions: Could Airports Really Become Greener?

Could airports really become greener? This book has tried to answer on this question by describing the processes of greening airports through the use of advanced technologies and operations. This includes concepts, strategies, and tactics for greening the entire air transport system, of which airports are an important component, together with airlines and ATC/ATM (Air Traffic Control/ Air Traffic Management). In particular, existing and prospective infrastructural, technical/technological, and operational performances of airports, ATC/ATM, and aircraft/airlines have been analysed regarding their impacts-costs on the environment and society in terms of consumption of non-renewable energy sources (aviation fuel as a derivative of crude oil) and related emissions of greenhouse gases and their prospective impacts on global warming. This additionally includes a detailed elaboration of the methodology for monitoring, analysing, and assessing the level of greening, i.e., sustainable development of airports. This methodology contains concepts, strategies, and—at its core—an indicator system for quantifying particular effects-benefits and impacts-costs over the medium-to long-term period. This indicator system consists of particular indicators and their measures reflecting the airport's infrastructural, technical/technological, operational, economic, environmental, and social performances while respecting the preferences of particular parties involved. Their estimation for particular cases has indicated that many airports have already began greening in relative terms, expressed by the generally decreasing quantities of particular impacts-costs per unit of their output-passengers, cargo shipments, and/or aircraft handled. However, in absolute terms, these impacts-costs have increased, albeit in many cases at decreasing rates. Such growth of particular impacts-costs reflects the endeavours of affected airports to appropriately accommodate air transport demand, which has been growing at higher annual rates than the technical/technological and operational improvements intended to mitigate them.

Developing airports into true multimodal transport nodes has proven to have substantive greening effects. Particularly in Europe, such development has implied connecting (including) some large hub airports to the medium-to long-distance

M. Janić, *Greening Airports*, Green Energy and Technology,
DOI: 10.1007/978-0-85729-658-0_8, © Springer-Verlag London Limited 2011

surface—usually HSR (High Speed Rail)—transport network. This has enabled substitution of some short-haul flights with equivalent HSR services either through modal competition or complementarity. Estimates related to the prospective inclusion of large hub airports into the HSR network have confirmed again that even on a modest scale, the above-mentioned substitution has substantial potential for mitigating the overall airport airside impacts-costs (externalities) in terms of airside congestion and delays of airlines and air passenger, local noise, and energy consumption and related emissions of greenhouse gases.

Further, two sets of distinctive advanced technologies and operations for greening, i.e., sustainable development of the airport airside and landside area, have been elaborated. Specifically, in the first set, the potential of two types of technologies have been considered for greening airports in their airside area. The first type includes innovative operational procedures supported by advanced technologies to increase airport runway capacity. These include ATC time-and vertical-distance-based separation rules, and prioritising aircraft landings on a single runway. Some estimates based on the "what-if" scenario approach have shown that they could all substantially increase the airport runway landing capacity, consequently mitigating airside congestion and delays, and thus postpone the need for building the new runways requiring additional (new) land. The other type of technologies relates to replacing conventional Jet A aviation fuel (kerosene) as a derivative of non-renewable crude oil with renewable LH_2 (Liquid Hydrogen) at both global (air transport system) and local (airport) scales. Estimations, again based on the "what-if" scenario approach, indicate that replacement of conventional aircraft with cryogenic aircraft even at very modest annual rates could contribute to slowing down, stagnating, and then diminishing cumulative emissions of greenhouse gases in the atmosphere by both the system and the given airport over the long-term future.

In the other set, a single presumably advanced technology and related operations in the airport landside area expected to stimulate more intensive use of public transport by air passengers and airport employees in the scope of the airport ground access systems/modes are elaborated. The objective was to investigate the potential of this technology—a LRRT (Light Rail Rapid Transit) system serving a large hub airport-to eventually mitigate the social and environmental impacts-costs of airport ground access systems/modes in terms of road congestion and delays, noise, energy consumption and related emissions of greenhouse gases, and traffic incidents/accidents. Estimations again based on the "what-if" scenario approach have shown the substantial potential of LRRT in mitigating almost all the above-mentioned impacts and consequently making savings in their costs (externalities). In addition, it has been shown that implementation of electric cars will only marginally compromise the above-mentioned savings achieved by the LRRT system.

In summary, the described advanced technologies and operations should be applied selectively, depending on the airport in question, to ensure their social-economic feasibility on the one hand and materialising of the expected greening effects on the other.

Author Biography

Dr. Milan Janić is a transport and traffic engineer and planner. Currently, he is a Senior Researcher and Research Program Leader at the Transport and Infrastructure Section of the OTB Research Institute for the Built Environment of Delft University of Technology (Delft, the Netherlands) and The Principal Scientist/Research Professor at the Faculty of Traffic and Transport Engineering of the University of Belgrade (Belgrade, Serbia). Previously, he kept the posts of Senior Researcher at Manchester Metropolitan University (Manchester, the UK), Loughborough University (Transport Studies Group) (Loughborough, the UK), and the Institute of Transport Research (Ljubljana, Slovenia).

Dr. Janić has been involved in transport-related research and planning projects at both national and international levels for almost 30 years. He has also published many scientific and academic transport-related papers, and presented many of them at national and international conferences dedicated to transport. In addition, he has had three books published: "Airport Analysis, Planning and Design: Demand, Capacity and Congestion" (2009), "The Sustainability of Air Transportation: A Quantitative Analysis and Assessment" (2007), and "Air Transport System Analysis and Modelling: Capacity, Quality of Services and Economics" (2000).

In addition, Dr. Janić is a member of NECTAR (Network on European Communications and Transport Activity Research), Air Transport Research Society (ATRS), the Airfield and Airspace Capacity and Delay Committee of Transportation Research Board (TRB) (Washington, DC, USA), and the Delft Aviation Centre (Delft University of Technology, Delft, the Netherlands).

M. Janić, *Greening Airports*, Green Energy and Technology,
DOI: 10.1007/978-0-85729-658-0, © Springer-Verlag London Limited 2011

Glossary

A ACI, accessibility, accident, advanced, AEA, AIRBUS, aircraft, airport, airside area, airspace, ATAG, Air Traffic Control (ATC), algorithm, alternative fuel, Amsterdam, analyzing, approach speed, area, arrival, assessing, assumption, atm, air pollution, avionics

B BAA, benefits, boeing, BTS

C Capacity, car, carbon dioxide coal, competition, complementarity, congestion, consumption, cost, cryogenic, criteria, crude oil

D Delay, demand, departure. Design, DFT

E EC, EEC, effects, efficiency, electric car, emissions, employees, engine, EPA, equipment, EU, externalities economic

F FAA, facilities, fleet, flight, forecasting, fuel, function

G Glide slope, global, globalization, gravitational acceleration, greening, greenhouse gases, growth, ground

H Headway, heathrow, HSR (Hugh Speed Rail), hub, hydrogen

I ICAO, ICE, IEA, impacts, incident, indicator, infrastructure, IPCC

J Jet

K Kerosene

L Landing, landside area, land use, line, liquid hydrogen. Light rail rapid transit, logistics, load factor, London, LTO cycle, LLC (Low Cost Carrier)

M Mach number, market, Maximum take-off weight, measure, methodology, mitigating model, modeling, modal split, monitoring, multimodal, multinomial logit model, multiplier

N NASA, natural gas, Newton, New York, node, noise, nitrogen oxides, nuclear

M. Janić, *Greening Airports*, Green Energy and Technology,
DOI: 10.1007/978-0-85729-658-0, © Springer-Verlag London Limited 2011

O Objectives, operations

P Passengers, payload, performance, planning, population, priority, productivity, procedures, profitability, propagation

Q Quality quantity, queues, queuuing, quota

R Range, rate, revenues, RITA, route, runway, rail, rules

S Safety, savings, Schiphol, separation, social, society, specific fuel consumption, substitution, storage, supply, sustainability, system

T Take-off, technique, technical, technologies, TGV, threshold, time, tons, trading-off, traffic, train, transport, TRB, Typolev

U User, utility

V Variable, vapour, vertical distance, volume

W Warming waste, wake vortex, water, weather, wind

Y Yield

Z Zurich

Printed by Publishers' Graphics LLC
CAMZ140309.20.06.19